Donald Greenspan
Numerical Solution of Ordinary Differential Equations

Related Titles

Sewell, G.
The Numerical Solution of Ordinary and Partial Differential Equations
approx. 352 pages
2005
Hardcover
ISBN 0-471-73580-9

Hunt, B. R., Lipsman, R. L., Osborn, J. E., Rosenberg, J. M.
Differential Equations with Matlab
295 pages
Softcover
ISBN 0-471-71812-2

Butcher, J.C.
Numerical Methods for Ordinary Differential Equations
440 pages
2003
Set
ISBN 0-470-86827-9

Markley, N.G.
Principles of Differential Equations
340 pages
2004
Hardcover
ISBN 0-471-64956-2

Donald Greenspan

Numerical Solution of Ordinary Differential Equations

for Classical, Relativistic and Nano Systems

WILEY-VCH Verlag GmbH & Co. KGaA

The Author

Donald Greenspan
University of Texas
Mathematics Dept.
Arlington, Texas 76019
USA

Cover
aktivComm GmbH, Weinheim

All books published by Wiley-VCH are carefully produced. Nevertheless, authors, editors, and publisher do not warrant the information contained in these books, including this book, to be free of errors. Readers are advised to keep in mind that statements, data, illustrations, procedural details or other items may inadvertently be inaccurate.

Library of Congress Card No.: applied for.

British Library Cataloging-in-Publication Data:
A catalogue record for this book is available from the British Library.

Bibliographic information published by Die Deutsche Bibliothek
Die Deutsche Bibliothek lists this publication in the Deutsche Nationalbibliografie; detailed bibliographic data is available in the Internet at <http://dnb.ddb.de>.

© 2006 WILEY-VCH Verlag GmbH & Co. KGaA, Weinheim

All rights reserved (including those of translation into other languages). No part of this book may be reproduced in any form – nor transmitted or translated into machine language without written permission from the publishers. Registered names, trademarks, etc. used in this book, even when not specifically marked as such, are not to be considered unprotected by law.

Printed in the Federal Republic of Germany
Printed on acid-free paper

Typesetting Uwe Krieg, Berlin
Printing Strauss GmbH, Mörlenbach
Binding Litges & Dopf Buchbinderei GmbH, Heppenheim

ISBN-13: 978-3-527-40610-4
ISBN-10: 3-527-40610-7

Contents

Preface *IX*

1 **Euler's Method** *1*
1.1 Introduction *1*
1.2 Euler's Method *1*
1.3 Convergence of Euler's Method* *5*
1.4 Remarks *8*
1.5 Exercises *9*

2 **Runge–Kutta Methods** *11*
2.1 Introduction *11*
2.2 A Runge–Kutta Formula *11*
2.3 Higher-Order Runge–Kutta Formulas *15*
2.4 Kutta's Fourth-Order Formula *22*
2.5 Kutta's Formulas for Systems of First-Order Equations *23*
2.6 Kutta's Formulas for Second-Order Differential Equations *26*
2.7 Application – The Nonlinear Pendulum *28*
2.8 Application – Impulsive Forces *31*
2.9 Exercises *34*

3 **The Method of Taylor Expansions** *37*
3.1 Introduction *37*
3.2 First-Order Problems *37*
3.3 Systems of First-Order Equations *40*
3.4 Second-Order Initial Value Problems *41*
3.5 Application – The van der Pol Oscillator *43*
3.6 Exercises *45*

4 Large Second-Order Systems with Application to Nano Systems 49
- 4.1 Introduction 49
- 4.2 The N-Body Problem 49
- 4.3 Classical Molecular Potentials 50
- 4.4 Molecular Mechanics 52
- 4.5 The Leap Frog Formulas 52
- 4.6 Equations of Motion for Argon Vapor 53
- 4.7 A Cavity Problem 54
- 4.8 Computational Considerations 56
- 4.9 Examples of Primary Vortex Generation 56
- 4.10 Examples of Turbulent Flow 59
- 4.11 Remark 61
- 4.12 Molecular Formulas for Air 62
- 4.13 A Cavity Problem 63
- 4.14 Initial Data 64
- 4.15 Examples of Primary Vortex Generation 65
- 4.16 Turbulent Flow 66
- 4.17 Colliding Microdrops of Water Vapor 70
- 4.18 Remarks 72
- 4.19 Exercises 74

5 Completely Conservative, Covariant Numerical Methodology 77
- 5.1 Introduction 77
- 5.2 Mathematical Considerations 77
- 5.3 Numerical Methodology 78
- 5.4 Conservation Laws 79
- 5.5 Covariance 82
- 5.6 Application – A Spinning Top on a Smooth Horizontal Plane 85
- 5.7 Application – Calogero and Toda Hamiltonian Systems 103
- 5.8 Remarks 108
- 5.9 Exercises 109

6 Instability 111
- 6.1 Introduction 111
- 6.2 Instability Analysis 111
- 6.3 Numerical Solution of Mildly Nonlinear Autonomous Systems 122
- 6.4 Exercises 130

7 Numerical Solution of Tridiagonal Linear Algebraic Systems and Related Nonlinear Systems 133
- 7.1 Introduction 133
- 7.2 Tridiagonal Systems 133

7.3	The Direct Method	*136*
7.4	The Newton–Lieberstein Method	*137*
7.5	Exercises	*140*

8 Approximate Solution of Boundary Value Problems *143*
8.1	Introduction	*143*
8.2	Approximate Differentiation	*143*
8.3	Numerical Solution of Boundary Value Problems Using Difference Equations	*144*
8.4	Upwind Differencing	*148*
8.5	Mildly Nonlinear Boundary Value Problems	*150*
8.6	Theoretical Support*	*152*
8.7	Application – Approximation of Airy Functions	*155*
8.8	Exercises	*156*

9 Special Relativistic Motion *159*
9.1	Introduction	*159*
9.2	Inertial Frames	*160*
9.3	The Lorentz Transformation	*161*
9.4	Rod Contraction and Time Dilation	*161*
9.5	Relativistic Particle Motion	*163*
9.6	Covariance	*163*
9.7	Particle Motion	*165*
9.8	Numerical Methodology	*166*
9.9	Relativistic Harmonic Oscillation	*169*
9.10	Computational Covariance	*170*
9.11	Remarks	*174*
9.12	Exercises	*175*

10 Special Topics *177*
10.1	Introduction	*177*
10.2	Solving Boundary Value Problems by Initial Value Techniques	*177*
10.3	Solving Initial Value Problems by Boundary Value Techniques	*178*
10.4	Predictor-Corrector Methods	*179*
10.5	Multistep Methods	*180*
10.6	Other Methods	*180*
10.7	Consistency*	*181*
10.8	Differential Eigenvalue Problems	*182*
10.9	Chaos*	*184*
10.10	Contact Mechanics	*184*

Appendix

A **Basic Matrix Operations** *187*

Solutions to Selected Exercises *191*

References *197*

Index *203*

Preface

The study and application of ordinary differential equations has been a major part of the history of mathematics. In recent years, new applications in such areas as molecular mechanics and nanophysics have simply added to their significance.

This book is intended to be used as either a handbook or a text for a one-semester, introductory course in the numerical solution of ordinary differential equations. Theory, methodology, intuition, and applications are interwoven throughout. The choice of methods is guided by applied, rather than theoretical, interests. Throughout, nonlinearity and determinism are emphasized.

Chapter 1 develops Euler's method and fundamental convergence theory. Chapter 2 develops Runge–Kutta formulas through the highest order available, that is, order 10. Chapter 3 develops Taylor expansion methodology of arbitrary orders. Chapter 4 develops conservative numerical methodology. Chapter 5 is concerned with very large systems of differential equations, such as those used in molecular mechanics. Chapter 6 studies practical aspects of instability. Chapters 7 and 8 are concerned with boundary value problems. Chapter 9 presents, in an entirely self-contained fashion, fundamentals of special relativistic dynamics, in which the differential equations and the related constraints are truly unique. Chapter 10 is a survey with references of the many other topics available in the literature.

Flexibility is incorporated by providing programs generically. Computer technology is in such a rapid state of growth that the use of a specific programming language can become outdated in a very short time. In addition, the individual who wishes to use a graphics routine is free to use whichever is most readily available to him or her.

Relatively difficult sections are marked with an asterisk and may be omitted without disturbing the book's continuity.

Finally, it should be noted that this book contains materials of interest in engineering and science which are not available elsewhere. For example, Chapter 5 develops numerical methodology which conserves exactly the same energy, linear momentum and angular momentum as does a conservative continuous system.

I wish to thank the Institute of Mathematics and Its Applications for permission to reproduce Table 2.2.

Donald Greenspan

Arlington, Texas, 2005

1
Euler's Method

1.1
Introduction

In this chapter, we will consider a numerical method for a basic initial value problem, that is, for

$$y' = F(x,y), \quad y(0) = \alpha. \tag{1.1}$$

We will use a simplistic numerical method called Euler's method. Because of the simplicity of both the problem and the method, the related theory is relatively transparent and will be provided in detail. Though we will not do so, the theory developed in this chapter does extend to the more advanced methods to be introduced later, but only with increased complexity.

With respect to (1.1), we assume that a unique solution exists, but that analytical attempts to construct it have failed.

1.2
Euler's Method

Consider the problem of approximating a continuous function $y = f(x)$ on $x \geq 0$ which satisfies the differential equation

$$y' = F(x,y) \tag{1.2}$$

on $x > 0$, and the initial condition

$$y(0) = \alpha, \tag{1.3}$$

in which α is a given constant. In 1768 (see the Collected Works of L. Euler, vols. 11 (1913), 12 (1914)), L. Euler developed a method to prove that the initial value problem (1.2), (1.3) had a solution. The method was numerical in nature and today it is implemented on modern computers and is called Euler's method. The basic idea is as follows. By the definition of a derivative,

$$y'(x) = \lim_{h \to 0} \frac{f(x+h) - f(x)}{h}. \tag{1.4}$$

For small $h > 0$, then, (1.4) implies that a reasonable difference quotient approximation for $y'(x)$ is

$$y'(x) = \frac{f(x+h) - f(x)}{h}. \tag{1.5}$$

Substitution of (1.5) into (1.2) yields the difference equation

$$\frac{f(x+h) - f(x)}{h} = F(x,y) \tag{1.6}$$

which approximates the differential equation (1.2). However, (1.6) can be rewritten as

$$f(x+h) = f(x) + hF(x,y)$$

or, equivalently, as

$$y(x+h) = y(x) + hF(x,y(x)), \tag{1.7}$$

which enables one to approximate $y(x+h)$ in terms of $y(x)$ and $F(x,y(x))$. Equation (1.7) is the cornerstone of Euler's method, which is described precisely as follows.

Since a computer cannot calculate indefinitely, let $x \geq 0$ be replaced by $0 \leq x \leq L$, in which L is a positive constant. The value of L is usually determined by the physics of the phenomenon under consideration. If the phenomenon occurs over a short period of time, then L can be chosen to be relatively small. If the phenomenon is long lasting, then L must be relatively large. In either case, L is a fixed, positive constant. The interval $0 \leq x \leq L$ is then divided into n equal parts, each of length h, by the points $x_i = ih$, $i = 0, 1, 2, \ldots, n$. The value $h = L/n$ is called the grid size. The points x_i are called grid points. Let $y_i = y(x_i)$, $i = 0, 1, 2, \ldots, n$, so that initial condition (1.3) implies $y_0 = \alpha$. Next, at each of the grid points $x_0, x_1, x_2, \ldots, x_{n-1}$, approximate the differential equation by (3.6) in the notation

$$\frac{y_{i+1} - y_i}{h} = F(x_i, y_i), \quad i = 0, 1, 2, \ldots, n-1, \tag{1.8}$$

or, in explicit recursive form

$$y_{i+1} = y_i + hF(x_i, y_i), \quad i = 0, 1, 2, \ldots, n-1. \tag{1.9}$$

Then, beginning with

$$y_0 = \alpha, \tag{1.10}$$

set $i = 0$ in (1.9) and determine y_1. Knowing y_1, set $i = 1$ in (1.9) and determine y_2. Knowing y_2, set $i = 2$ in (1.9) and determine y_3, and so forth, until, finally, y_n is generated. The resulting discrete function $y_0, y_1, y_2, \ldots, y_n$ is called the numerical solution.

Example 1.1 Consider the initial value problem

$$y' + y = x, \quad y(0) = 1. \tag{1.11}$$

This is a linear problem and can be solved exactly to yield the solution

$$Y(x) = x - 1 + 2e^{-x}. \tag{1.12}$$

Hence, there is no need to solve (1.11) numerically. We will proceed numerically for illustrative purposes only. For Euler's method, fix $L = 1$ and $h = 0.2$. Then, $x_0 = 0.0$, $x_1 = 0.2$, $x_2 = 0.4$, $x_3 = 0.6$, $x_4 = 0.8$, $x_5 = 1.0$ and differential equation (1.11) is approximated by the difference equation

$$\frac{y_{i+1} - y_i}{0.2} + y_i = x_i, \quad i = 0, 1, 2, 3, 4,$$

or, equivalently, by

$$y_{i+1} = (0.8)y_i + (0.2)x_i, \quad i = 0, 1, 2, 3, 4. \tag{1.13}$$

Since $y_0 = 1$, (1.13) yields, to three decimal places

$$y_1 = (0.8)y_0 + (0.2)x_0 = (0.8)(1.000) + (0.2)(0.0) = 0.800$$
$$y_2 = (0.8)y_1 + (0.2)x_1 = (0.8)(0.800) + (0.2)(0.2) = 0.680$$
$$y_3 = (0.8)y_2 + (0.2)x_2 = (0.8)(0.680) + (0.2)(0.4) = 0.624$$
$$y_4 = (0.8)y_3 + (0.2)x_3 = (0.8)(0.624) + (0.2)(0.6) = 0.619$$
$$y_5 = (0.8)y_4 + (0.2)x_4 = (0.8)(0.619) + (0.2)(0.8) = 0.655.$$

Thus, the numerical approximation with $h = 0.2$ is

$$y(0.0) = 1.000$$
$$y(0.2) = 0.800$$
$$y(0.4) = 0.680$$
$$y(0.6) = 0.624$$
$$y(0.8) = 0.619$$
$$y(1.0) = 0.655.$$

However, from (1.12), the exact solution, rounded to three decimal places, at the grid points is given by

$$Y(0.0) = 1.000$$
$$Y(0.2) = 0.837$$
$$Y(0.4) = 0.741$$
$$Y(0.6) = 0.698$$
$$Y(0.8) = 0.699$$
$$Y(1.0) = 0.736.$$

Comparison of the numerical and the exact solutions then yields the precise amount of error that results at each grid point when employing Euler's method.

Now, unlike the above example, numerical methodology will be applied only when the exact solution of (1.2), (1.3) is *not* known. Thus, in practice the error at each grid point will not be known. It is essential then to know, *a priori*, that the unknown error at each grid point is arbitrarily small if h is arbitrarily small, that is, that the error at each grid point decreases to zero as h decreases to zero. If this were valid, then one would have the assurance that the error generated by Euler's method is negligible for all sufficiently small grid sizes h. That this is correct *when all calculations are exact* will be established next.

A generic algorithm for Euler's method is given as follows.

Algorithm 1 Euler

Step 1. Set a counter $k = 1$.
Step 2. Set a time step h.
Step 3. Set an initial time x.
Step 4. Set initial value y.
Step 5. Calculate
$$K_0 = y$$
$$K_1 = hF(x,y).$$
Step 6. Calculate y at $x + h$ by
$$y(x+h) = (K_0 + K_1).$$
Step 7. Increase the counter from k to $k+1$.
Step 8. Set $y = y(x+h)$, $x = x + h$.
Step 9. Repeat Steps 5–8.
Step 10. Continue until $k = 100$.

1.3
Convergence of Euler's Method*

We wish to show now that, for Euler's method, *the error at each grid point decreases to zero as h decreases to zero*. The associated theory is called convergence theory. In developing convergence theory, we will require some preliminary results.

Lemma 1.1 *If the numbers $|E_i|$, $i = 0, 1, 2, 3, \ldots, n$, satisfy*

$$|E_{i+1}| \leq A |E_i| + B, \quad i = 0, 1, 2, 3, \ldots, n-1 \tag{1.14}$$

where A and B are nonnegative constants and $A \neq 1$, then

$$|E_i| \leq A^i |E_0| + \frac{A^i - 1}{A - 1} B, \quad i = 1, 2, 3 \ldots, n \tag{1.15}$$

Proof. For $i = 0$, (1.14) yields

$$|E_1| \leq A |E_0| + B = A |E_0| + \frac{A - 1}{A - 1} B,$$

so that (1.15) is valid for $i = 1$. The proof is now completed by induction. Assume that for fixed i, (1.15) is valid, that is,

$$|E_i| \leq A^i |E_0| + \frac{A^i - 1}{A - 1} B.$$

Then we must prove that

$$|E_{i+1}| \leq A^{i+1} |E_0| + \frac{A^{i+1} - 1}{A - 1} B.$$

Since, by (1.14),

$$|E_{i+1}| \leq A |E_i| + B,$$

then

$$|E_{i+1}| \leq A \left[A^i |E_0| + \frac{A^i - 1}{A - 1} B \right] + B = A^{i+1} |E_0| + \frac{A^{i+1} - 1}{A - 1} B,$$

and the proof is complete. $\square$

The value of Lemma 1.1 is as follows. If each term of a sequence $|E_0|, |E_1|, |E_2|, |E_3|, |E_4|, \ldots, |E_n|, \ldots$, is related to the previous term by (1.14), then Lemma 1.1 enables one to relate each term directly to $|E_0|$ only, that is, to the very first term of the sequence.

Theorem 1.1 Let I be the open interval $0 < x < L$ and $\bar{I}$ the closed interval $0 \leq x \leq L$. Assume the initial value problem

$$y' = F(x,y), \quad y(0) = \alpha \tag{1.16}$$

has the unique solution $Y(x)$ on $\bar{I}$. Then, on I,

$$Y'(x) \equiv F(x, Y(x)) \tag{1.17}$$

and

$$Y(0) = \alpha. \tag{1.18}$$

Assume that $Y'(x)$ and $Y''(x)$ are continuous and that there exist positive constants M, N such that

$$|Y''(x)| \leq N, \quad 0 \leq x \leq L \tag{1.19}$$

$$\left|\frac{\partial F}{\partial y}\right| \leq M, \quad 0 \leq x \leq L, \quad -\infty < y < \infty. \tag{1.20}$$

Next, let $\bar{I}$ be subdivided into n equal parts by the grid points $x_0 < x_1 < x_2 < \ldots < x_n$, where $x_0 = 0$, $x_n = L$. The grid size h is given by

$$h = L/n. \tag{1.21}$$

Let y_k be the numerical solution of (1.16) by Euler's method on the grid points, so that

$$y_{k+1} = y_k + hF(x_k, y_k), \quad k = 0, 1, 2, \ldots, n-1 \tag{1.22}$$

$$y_0 = \alpha. \tag{1.23}$$

Finally, define the error E_k at each grid point x_k by

$$E_k = Y_k - y_k, \quad k = 0, 1, 2, 3, \ldots, n. \tag{1.24}$$

Then,

$$|E_k| \leq \frac{\left[(1+Mh)^k - 1\right] Nh}{2M}, \quad k = 0, 1, 2, 3, \ldots, n. \tag{1.25}$$

Proof. Consider

$$|E_{k+1}| = |Y_{k+1} - y_{k+1}|.$$

Then

$$|E_{k+1}| = |Y_{k+1} - y_{k+1}| = |Y(x_k + h) - (y_k + hF(x_k, y_k))|.$$

Introducing a Taylor expansion for $Y(x_k + h)$ implies

$$|E_{k+1}| = \left|\left(Y(x_k) + hY'(x_k) + \frac{1}{2}h^2 Y''(\xi)\right) - (y_k + hF(x_k, y_k))\right|$$

$$= \left|Y_k - y_k + h\left[Y'(x_k) - F(x_k, y_k)\right] + \frac{1}{2}h^2 Y''(\xi)\right|.$$

From (1.17), then

$$|E_{k+1}| = \left|Y_k - y_k + h\left[F(x_k, Y_k) - F(x_k, y_k)\right] + \frac{1}{2}h^2 Y''(\xi)\right|,$$

which, by the mean value theorem for a function of two variables, implies

$$|E_{k+1}| = \left|Y_k - y_k + h\left[(Y_k - y_k)\frac{\partial F}{\partial y}(x_k, \eta)\right] + \frac{1}{2}h^2 Y''(\xi)\right|$$

$$= \left|(Y_k - y_k)\left(1 + h\frac{\partial F}{\partial y}\right) + \frac{1}{2}h^2 Y''(\xi)\right|.$$

Hence, by the rules for absolute values,

$$|E_{k+1}| \leq |Y_k - y_k|\left(1 + h\left|\frac{\partial F}{\partial y}\right|\right) + \frac{1}{2}h^2 |Y''(\xi)|,$$

which, by (1.19), (1.20) yields

$$|E_{k+1}| \leq |Y_k - y_k|(1 + Mh) + \frac{1}{2}h^2 N.$$

Thus, since $|Y_k - y_k| = |E_k|$, one has

$$|E_{k+1}| \leq |E_k|(1 + Mh) + \frac{1}{2}h^2 N. \tag{1.26}$$

Application of Lemma 1.1 to (1.26) with $A = (1 + Mh)$, $B = \frac{1}{2}h^2 N$ then implies

$$|E_k| \leq (1 + Mh)^k |E_0| + \frac{(1 + Mh)^k - 1}{(1 + Mh) - 1}\left(\frac{1}{2}h^2 N\right). \tag{1.27}$$

However, since $Y(0) = y(0) = \alpha$, one has $E_0 = 0$, so that (1.27) simplifies to

$$|E_k| \leq \frac{\left[(1 + Mh)^k - 1\right] Nh}{2M}, \quad k = 0, 1, 2, 3, \ldots, n, \tag{1.28}$$

and the theorem is proved. □

Theorem 1.2 *Under the assumptions of Theorem 1.1, one has that at each grid point*

$$\lim_{h \to 0} |E_k| = 0, \quad k = 0, 1, 2, 3, \ldots, n.$$

Proof. Since $(1 + Mh) > 1$, the largest value of $(1 + Mh)^k$ results when $k = n$. Thus, from (1.28),

$$|E_k| \leq \frac{[(1 + Mh)^n - 1] Nh}{2M}, \tag{1.29}$$

which, by (1.21), implies

$$|E_k| \le \frac{\left[(1+Mh)^{L/h} - 1\right] Nh}{2M}. \tag{1.30}$$

By the laws of exponents, then,

$$|E_k| \le \frac{\left\{\left[(1+Mh)^{\frac{1}{Mh}}\right]^{ML} - 1\right\} Nh}{2M}. \tag{1.31}$$

Note now that if $Mh = \gamma$, then

$$\lim_{h\to 0} Mh = \lim_{\gamma \to 0} \gamma = 0.$$

Thus,

$$\lim_{h \to 0}\left[(1+Mh)^{\frac{1}{Mh}}\right]^{ML} = \lim_{\gamma \to 0}\left[(1+\gamma)^{\frac{1}{\gamma}}\right]^{ML}.$$

But, $\lim_{\gamma \to 0}\left[(1+\gamma)^{\frac{1}{\gamma}}\right] = e$. Thus,

$$\lim_{h \to 0} \frac{\left\{\left[(1+Mh)^{\frac{1}{Mh}}\right]^{ML} - 1\right\} Nh}{2M} = \lim_{h \to 0} \frac{\{e^{ML} - 1\} Nh}{2M} = 0.$$

Thus, from (1.31), $\lim_{h \to 0} |E_k| = 0$ for all values of k, and the theorem is proved. $\square$

1.4
Remarks

In practice, as will be shown soon, numerical methods which are more economical and more accurate than Euler's method can be developed easily. However, convergence proofs for these methods are more complex than for Euler's method.

Note that the essence of Theorem 1.2 is that if one wishes arbitrarily high accuracy, one need only choose h sufficiently small. Unfortunately, such remarks are purely qualitative. Indeed, if one has a prescribed accuracy, Theorems 1.1 and 1.2 do not allow one to determine the precise h, *a priori*, since the constant N in (1.19) is rarely known exactly and the practical matter of roundoff error in actual calculations has not been included in the theorems. The determination of accuracy is often estimated in an *a posteriori* manner as follows. One calculates for both h and $\frac{1}{2}h$ and takes those figures which are in agreement for the two calculations. For example, if at a point x and for $h = 0.1$ one finds

$y = 0.876\,532$ while for $h = 0.05$ one finds at the same point that $y = 0.876\,513$, then one assumes that the result $y = 0.8765$ is *an* accurate result.

As noted above, Theorems 1.1 and 1.2 do not consider roundoff error, which is always present in computer calculations. At the present time there is no universally accepted method to analyze roundoff error after a large number of time steps. The three main methods for analyzing roundoff accumulation are the analytical method (Henrici (1962), (1963)), the probabilistic method (Henrici (1962), (1963)) and the interval arithmetic method (Moore (1979)), each of which has both advantages and disadvantages.

1.5
Exercises

1.1 With $h = 0.1$, find the numerical solution on $0 \leq x \leq 1$ by Euler's method for
$$y' = y^2 + 2x - x^4, \quad y(0) = 0.$$
and compare your results with the exact solution $y = x^2$.

1.2 With $h = 0.1$, find the numerical solution on $0 \leq x \leq 2$ by Euler's method for
$$y' = y^3 - 8x^3 + 2, \quad y(0) = 0$$
and compare your results with the exact solution $y = 2x$.

1.3 With $h = 0.05$, find the numerical solution on $0 \leq x \leq 1$ by Euler's method for
$$y' = xy^2 - 2y, \quad y(0) = 1.$$
Find the exact solution and compare the numerical results with it.

1.4 With $h = 0.01$, find the numerical solution on $0 \leq x \leq 2$ by Euler's method for
$$y' = -2xy^2, \quad y(0) = 1,$$
and compare your results with the exact solution $y = \frac{1}{1+x^2}$.

1.5 With $h = 0.05$, find the numerical solution on $0 \leq x \leq 1$ by Euler's method for
$$y' = e^y - e^{x^2} + 2x, \quad y(0) = 0,$$
and compare your results with the exact solution $y = x^2$.

1.6 With $h = 0.01$, find the numerical solution on $0 \leq x \leq 10$ by Euler's method for
$$y' = y - \frac{2+x}{(1+x)^2}, \quad y(0) = 1$$
and compare your results with the exact solution $y = \frac{1}{1+x}$.

1.7 Estimate the value M in Theorem 1.1 for each of the following. If possible, also estimate the value of N.

(a) $y' = x + \sin y$, $\quad 0 < x < 1$
(b) $y' = x^2 \cos y$, $\quad 0 < x < 2$
(c) $y' = x + y$, $\quad 0 < x < 3$.

2
Runge–Kutta Methods

2.1
Introduction

Let us show now how, for the same grid size h, one can replace Euler's formula by a relatively simple formula which will yield improved accuracy. The formulas to be developed in this chapter are called Runge–Kutta formulas because the initial work was done by C. Runge and W. Kutta around 1900 (Butcher (1987); Kutta (1901); Runge (1895)). The basic idea is to use information which is available, but has been neglected by the Euler formula. Indeed, for the equation

$$y' = F(x,y), \qquad (2.1)$$

the Euler formula approximates y_{k+1} by evaluating $F(x_k, y_k)$, that is, F is evaluated at the single point (x_k, y_k). But, $F(x, y)$ is known at every point in the (x, y) plane. Runge–Kutta formulas make use of this fact, as is shown next.

2.2
A Runge–Kutta Formula

Consider the following difference approximation of (2.1) at the grid point x_i:

$$\frac{y_{i+1} - y_i}{h} = \alpha F(x_i, y_i) + \beta F(x_i + \gamma h, y_i + \delta h), \qquad (2.2)$$

in which α, β, γ, δ are parameters to be determined. Of course, if one sets $\alpha = 1$ and $\beta = 0$, then Euler's formula results. Let us now rewrite (2.2) in the equivalent form

$$y_{i+1} = y_i + \alpha h F(x_i, y_i) + \beta h F(x_i + \gamma h, y_i + \delta h). \qquad (2.3)$$

Numerical Solution of Ordinary Differential Equations for Classical, Relativistic and Nano Systems. Donald Greenspan
Copyright © 2006 WILEY-VCH Verlag GmbH & Co. KGaA, Weinheim
ISBN: 3-527-40610-7

Using a Taylor expansion and the notation

$$F_x = \frac{\partial F}{\partial x},\quad F_y = \frac{\partial F}{\partial y},\quad F_{xx} = \frac{\partial^2 F}{\partial x^2},\quad F_{xy} = \frac{\partial^2 F}{\partial x \partial y},\quad F_{yy} = \frac{\partial^2 F}{\partial y^2},\ \ldots.$$

we have

$$\begin{aligned}F(x_i + \gamma h, y_i + \delta h) &= F(x_i, y_i) + \gamma h F_x(x_i, y_i) + \delta h F_y(x_i, y_i)\\ &\quad + \frac{1}{2}\left[\gamma^2 h^2 F_{xx}(x_i, y_i) + 2\gamma\delta h^2 F_{xy}(x_i, y_i) + \delta^2 h^2 F_{yy}(x_i, y_i)\right] \quad (2.4)\\ &\quad + O(h^3).\end{aligned}$$

Substitution of (2.4) into (2.3) and recombination yields

$$\begin{aligned}y_{i+1} &= y_i + h(\alpha + \beta) F(x_i, y_i) + \frac{1}{2}h^2\left[2\beta\gamma F_x(x_i,y_i) + 2\beta\delta F_y(x_i y_i)\right]\\ &\quad + \frac{h^3}{6}\left[3\beta\gamma^2 F_{xx}(x_i, y_i) + 6\beta\gamma\delta F_{xy}(x_i, y_i) + 3\beta\delta^2 F_{yy}(x_i, y_i)\right] \quad (2.5)\\ &\quad + \beta O(h^4).\end{aligned}$$

Note that (2.5) is of the particular form

$$y_{i+1} = y_i + h F_1 + \frac{1}{2}h^2 F_2 + \frac{h^3}{6}F_3 + O(h^4), \quad (2.6)$$

in which F_1, F_2, F_3 are the respective coefficients in (2.5). Note also that (2.6) is suggestive of a Taylor expansion for a function of one variable. So, suppose next that an exact solution of (2.1) is $Y(x)$ and that $Y(x)$ has the Taylor expansion

$$Y_{i+1} = Y_i + h Y_i' + \frac{1}{2}h^2 Y_i'' + \frac{h^3}{6}Y_i''' + \ldots + R_n. \quad (2.7)$$

From

$$Y' \equiv F(x, Y)$$

it follows from the chain rule and by implicit differentiation that

$$\begin{aligned}Y'' &\equiv F_x + F_y F\\ Y''' &\equiv F_{xx} + 2F F_{xy} + F^2 F_{yy} + F_x F_y + F(F_y)^2,\end{aligned}$$

which, upon substitution into (2.7), yields

$$\begin{aligned}Y_{i+1} &= Y_i + h F(x_i, Y_i) + \frac{1}{2}h^2\left\{F_x(x_i, Y_i) + F_y(x_i, Y_i) F(x_i, Y_i)\right\}\\ &\quad + \frac{h^3}{6}\left\{F_{xx}(x_i, Y_i) + 2F(x_i, Y_i) F_{xy}(x_i, Y_i) + [F(x_i, Y_i)]^2 F_{yy}(x_i, Y_i)\right. \quad (2.8)\\ &\quad \left. + F_x(x_i, Y_i) F_y(x_i, Y_i) + F(x_i, Y_i)\left[F_y(x_i, Y_i)\right]^2\right\} + O(h^4).\end{aligned}$$

2.2 A Runge–Kutta Formula

Now, (2.5) is an expansion for the numerical solution, while (2.8) is an expansion for the exact solution. Let us then explore the ways in which one can choose $\alpha, \beta, \gamma, \delta$ so that if $y_i = Y_i$, then (2.5) for y_{i+1} agrees in some sense with (2.8) for Y_{i+1}.

For the special choices $\alpha = 1$, $\beta = 0$, (2.5) agrees with (2.8) up to, but not including, the h^2 terms. This choice yields, of course, Euler's formula. If we wish to construct a formula for which (2.5) agrees with (2.8) through, at least, the h^2 terms, and in this sense is a better approximation than Euler's formula, we need only set the corresponding coefficients of the h and the h^2 terms equal, so that

$$\alpha + \beta = 1$$
$$2\beta\gamma F_x + 2\beta\delta F_y = F_x + FF_y,$$

which is a system of two equations in the four parameters $\alpha, \beta, \gamma, \delta$. Since four parameters are involved, we can, without penalty, expand the above system of two equations to the following three equations:

$$\alpha + \beta = 1$$
$$2\beta\gamma = 1$$
$$2\beta\delta = F$$

by requiring that the corresponding coefficients of F_x and F_y also be equal. This system of three equations in the four parameters has an infinite number of solutions, a simple and convenient one being

$$\alpha = \beta = \frac{1}{2}, \quad \gamma = 1, \quad \delta = F(x_i, y_i). \tag{2.9}$$

Substitution of (2.9) into (2.3) then yields

$$y_{i+1} = y_i + \frac{1}{2}hF(x_i, y_i) + \frac{1}{2}hF(x_{i+1}, y_i + hF(x_i, y_i)), \tag{2.10}$$

which is called a second-order Runge–Kutta formula, because the Taylor expansions (2.6) and (2.7) are identical through the terms of order h^2 under the assumption that $y_i = Y_i$. The formula (2.10) is "better" than Euler's formula because Euler's formula yields a value y_{i+1} which agrees with Y_{i+1} only through the terms of order h in the respective Taylor expansions.

The terms of the exact Taylor expansion with which the numerical solution does not agree are called the truncation error terms, or, simply, the *truncation error*.

Calculation with (2.10) can be done efficiently with a three-step algorithm which can be derived as follows. First, let

$$K_0 = hF(x_i, y_i).$$

Then, (2.10) becomes

$$y_{i+1} = y_i + \frac{1}{2}hK_0 + \frac{1}{2}hF(x_{i+1}, y_i + K_0). \tag{2.11}$$

Next, let

$$K_1 = hF(x_{i+1}, y_i + K_0),$$

so that (2.11) becomes

$$y_{i+1} = y_i + \frac{1}{2}(K_0 + K_1).$$

Euler's method is then summarized as follows. For each of $i = 0, 1, 2, 3, \ldots, n-1$, determine y_{i+1} from the three-step calculation

$$K_0 = hF(x_i, y_i) \tag{2.12}$$
$$K_1 = hF(x_{i+1}, y_i + K_0) \tag{2.13}$$
$$y_{i+1} = y_i + \frac{1}{2}(K_0 + K_1). \tag{2.14}$$

Example 2.1 Consider again the initial value problem

$$y' + y = x, \quad y(0) = 1, \tag{2.15}$$

which is rewritten as

$$y' = x - y, \quad y(0) = 1, \tag{2.16}$$

so that

$$F(x, y) = x - y. \tag{2.17}$$

Let us solve (2.16) on $0 \leq x \leq 1$ using (2.12)–(2.14) with $h = 0.1$. Since $h = 0.1$, (2.12)–(2.14) become

$$K_0 = 0.1(-y_i + x_i)$$
$$K_1 = 0.1(-y_i - K_0 + x_{i+1})$$
$$y_{i+1} = y_i + \frac{1}{2}(K_0 + K_1).$$

Thus, for $i = 0$, we find

$$K_0 = 0.1(-y_0 + x_0) = 0.1(-1) = -0.100$$
$$K_1 = 0.1(-y_0 - K_0 + x_1) = 0.1(-1 + 0.1 + 0.1) = -0.080 \tag{2.18}$$
$$y_1 = y_0 + \frac{1}{2}(K_0 + K_1) = 1 + \frac{1}{2}(-0.1 - 0.08) = 0.910.$$

Next, for $i = 1$, we have

$$K_0 = 0.1(-y_1 + x_1) = 0.1(-0.910 + 0.1) = -0.081$$
$$K_1 = 0.1(-y_1 - K_0 + x_2) = 0.1(-0.910 + 0.081 + 0.2) = -0.0629 \tag{2.19}$$
$$y_2 = 0.910 + \frac{1}{2}(-0.081 - 0.0629) = 0.838.$$

Continuing in the indicated fashion, we find, to three decimal places

$$y_3 = 0.782, \quad y_4 = 0.742, \quad y_5 = 0.714, \quad y_6 = 0.699$$
$$y_7 = 0.694, \quad y_8 = 0.700, \quad y_9 = 0.714, \quad y_{10} = 0.737.$$

For purposes of comparison, we have listed to three decimal places in Table 2.1, the exact solution $Y = x - 1 + 2e^{-x}$, the numerical solution by Runge–Kutta formula (2.10), and the numerical solution by Euler's method. The superiority of formula (2.10) is readily apparent.

Table 2.1 Comparison of the exact solution of $Y = x - 1$ ·'
Runge–Kutta (R-K) and Euler method numerical solut¨

	Exact	Euler	R-K
$x_0 = 0.0$	1.000	1.000	1ᴖ
$x_1 = 0.1$	0.910	0.900	
$x_2 = 0.2$	0.837	0 ᵒ	
$x_3 = 0.3$	0.782		
$x_4 = 0.4$	0.7´		
$x_5 = 0.5$			
$x_6 = 0.6$			
$x_7 = 0.7$	0.		
$x_8 = 0.8$	0.6		
$x_9 = 0.9$	0.71.		
$x_{10} = 1.0$	0.736		

2.3
Higher-Order Runge–Kutta ⌐ıas

Since the choice of α, β, γ, δ in (2.9) is in no way unique, one can construct a variety of second-order Runge–Kutta formulas. The same remark is valid for Runge–Kutta formulas of orders 3–10. In this section we will record at least one such popular formula for various orders. It should be noted that no Runge–Kutta formula of order 11 is available at present.

A third-order formula of Heun (1900) is given by

$$y_{i+1} = y_i + 0.25(K_0 + 3K_2) \tag{2.20}$$

in which

$$K_0 = hF(x_i, y_i)$$
$$K_1 = hF\left(x_i + \frac{h}{3}, y_i + \frac{K_0}{3}\right)$$
$$K_2 = hF\left(x_i + \frac{2h}{3}, y_i + \frac{2K_1}{3}\right)$$

By far the most popular formula is the Kutta fourth-order formula (Kutta (1901)):

$$y_{i+1} = y_i + \frac{1}{6}(K_0 + 2K_1 + 2K_2 + K_3) \tag{2.21}$$

in which

$$K_0 = hF(x_i, y_i)$$
$$K_1 = hF\left(x_i + \frac{1}{2}h, y_i + \frac{1}{2}K_0\right)$$
$$K_2 = hF\left(x_i + \frac{1}{2}h, y_i + \frac{1}{2}K_1\right)$$
$$K_3 = hF(x_{i+1}, y_i + K_2).$$

A relatively popular fifth-order formula is that of Nystrom (Nystrom (1925); Jain (1984)):

$$y_{i+1} = y_i + (23K_1 + 125K_2 - 81K_5 + 125K_6)/192 \tag{2.22}$$

in which

$$K_1 = hF(x_i, y_i)$$
$$K_2 = hF\left(x_i + \frac{1}{3}h, y_i + \frac{1}{3}K_1\right)$$
$$K_3 = hF\left(x_i + \frac{2}{5}h, y_i + \frac{(4K_1 + 6K_2)}{25}\right)$$
$$K_4 = hF\left(x_i + h, y_i + \frac{(K_1 - 12K_2 + 15K_3)}{4}\right)$$
$$K_5 = hF\left(x_i + \frac{2}{3}h, y_i + \frac{(6K_1 + 90K_2 - 50K_3 + 8K_4)}{81}\right)$$
$$K_6 = hF\left(x_i + \frac{4}{5}h, y_i + \frac{(6K_1 + 36K_2 + 10K_3 + 8K_4)}{75}\right).$$

Of more recent origin are the remarkable continuous Runge–Kutta formulas of Sarafyan ((1984), (1990)), one of fifth-order being given by

$$y_{i+c} = y_i + K_0 c + (-89K_0 + 96K_2 + 36K_3 - 64K_4 + 21K_5)\frac{c^2}{30}$$
$$+ \frac{2}{45}(71K_0 - 104K_2 - 54K_3 + 136K_4 - 49K_5)c^3 \tag{2.23}$$
$$+ \frac{2}{9}(-5K_0 + 8K_2 + 6K_3 - 16K_4 + 7K_5)c^4$$

in which

$$K_0 = hF(x_i, y_i)$$
$$K_1 = hF\left(x_i + \frac{1}{6}h, y_i + \frac{1}{6}K_0\right)$$
$$K_2 = hF\left(x_i + \frac{1}{4}h, y_i + \frac{1}{16}(K_0 + 3K_1)\right)$$
$$K_3 = hF\left(x_i + \frac{1}{2}h, y_i + \frac{1}{4}(K_0 - 3K_1 + 4K_2)\right)$$
$$K_4 = hF\left(x_i + \frac{3}{4}h, y_i + \frac{1}{16}(3K_0 + 9K_3)\right)$$
$$K_5 = hF\left(x_i + h, y_i + \frac{1}{7}(-4K_0 + 3K_1 + 12K_2 - 12K_3 + 8K_4)\right)$$

and

$$y_{i+c} = y(x_i + ch), \quad 0 \le c \le 1.$$

Formula (2.23) is not only fifth order at the grid points but is fourth order at each point between the grid points.

The sixth-order formula of Butcher (Jain (1984)) is:

$$y_{i+1} = y_i + \frac{1}{120}(11K_0 + 81K_2 + 81K_3 - 32K_4 - 32K_5 + 11K_6) \quad (2.24)$$

where

$$K_0 = hF(x_i, y_i)$$
$$K_1 = hF\left(x_i + \frac{1}{3}h, y_i + \frac{1}{3}K_0\right)$$
$$K_2 = hF\left(x_i + \frac{2}{3}h, y_i + \frac{2}{3}K_1\right)$$
$$K_3 = hF\left(x_i + \frac{1}{3}h, y_i + \frac{1}{12}K_0 + \frac{1}{3}K_1 - \frac{1}{2}K_2\right)$$
$$K_4 = hF\left(x_i + \frac{1}{2}h, y_i - \frac{1}{16}K_0 + \frac{9}{8}K_1 + \frac{3}{16}K_2 - \frac{3}{8}K_3\right)$$
$$K_5 = hF\left(x_i + \frac{1}{2}h, y_i + \frac{9}{8}K_1 - \frac{3}{8}K_2 - \frac{3}{4}K_3 + \frac{1}{2}K_4\right)$$
$$K_6 = hF\left(x_i + h, y_i + \frac{9}{44}K_0 - \frac{9}{11}K_1 + \frac{63}{44}K_2 + \frac{18}{11}K_3 - \frac{16}{11}K_5\right).$$

A seventh-order formula of Sarafyan (1970) is

$$y_{i+1} = y_i + \frac{1}{840}[41(K_0 + K_9) + 216(K_4 + K_8) + 27(K_5 + K_7) + 272K_6] \quad (2.25)$$

in which

$$K_0 = hF(x_i, y_i)$$

$$K_1 = hF\left(x_i + \frac{1}{3}h, y_i + \frac{1}{3}K_0\right)$$

$$K_2 = hF\left(x_i + \frac{1}{2}h, y_i + \frac{1}{8}[K_0 + 3K_1]\right)$$

$$K_3 = hF\left(x_i + h, y_i + \frac{1}{2}[K_0 - 3K_1 + 4K_2]\right)$$

$$K_4 = hF\left(x_i + \frac{1}{6}h, y_i + \frac{1}{648}[83K_0 + 32K_2 - 7K_3]\right)$$

$$K_5 = hF\left(x_i + \frac{1}{3}h, y_i + \frac{1}{54}[-3K_0 - 4K_2 + K_3 + 24K_4]\right)$$

$$K_6 = hF\left(x_i + \frac{1}{2}h, y_i + \frac{1}{5088}[-290K_0 - 524K_2 + 145K_3 + 1908K_4 + 1305K_5]\right)$$

$$K_7 = hF\left(x_i + \frac{2}{3}h, y_i + \frac{1}{1431}[292K_0 + 108K_2 + 13K_3 - 318K_4 + 753K_5 + 106K_6]\right)$$

$$K_8 = hF\left(x_i + \frac{5}{6}h, y_i + \frac{1}{68688}[14042K_0 + 11012K_2 - 4477K_3 + 5724K_4 - 6903K_5 + 6360K_6 + 31482K_7]\right)$$

$$K_9 = hF\left(x_i + h, y_i + \frac{1}{4346}[-2049K_0 - 1836K_2 + 839K_3 + 5724K_4 - 4692K_5 + 12084K_6 - 9540K_7 + 3816K_8]\right)$$

An eighth-order formula due to Shanks (1966) is

$$y_{i+1} = y_i + h\left(\frac{41}{840}K_0 + \frac{9}{35}K_5 + \frac{34}{105}K_6 + \frac{9}{280}K_7 + \frac{9}{280}K_8 + \frac{3}{70}K_9 + \frac{3}{14}K_{10} + \frac{41}{840}K_{11}\right) \quad (2.26)$$

in which

$$K_0 = F(x_i, y_i)$$

$$K_1 = F\left(x_i + \frac{1}{9}h, y_i + \frac{1}{9}hK_0\right)$$

2.3 Higher-Order Runge–Kutta Formulas

$$K_2 = F\left(x_i + \frac{1}{6}h, y_i + \frac{1}{24}h(K_0 + 3K_1)\right)$$

$$K_3 = F\left(x_i + \frac{1}{4}h, y_i + \frac{1}{16}h(K_0 + 3K_2)\right)$$

$$K_4 = F\left(x_i + \frac{1}{10}h, y_i + \frac{1}{500}h(29K_0 + 33K_2 - 12K_3)\right)$$

$$K_5 = F\left(x_i + \frac{1}{6}h, y_i + \frac{1}{972}h(33K_0 + 4K_3 + 125K_4)\right)$$

$$K_6 = F\left(x_i + \frac{1}{2}h, y_i + \frac{1}{36}h(-21K_0 + 76K_3 + 125K_4 - 162K_5)\right)$$

$$K_7 = F\left(x_i + \frac{2}{3}h, y_i + \frac{1}{243}h(-30K_0 - 32K_3 + 125K_4 + 99K_6)\right)$$

$$K_8 = F\left(x_i + \frac{1}{3}h, y_i + \frac{1}{324}h(1175K_0 - 3456K_3 - 6250K_4 + 8424K_5 \right.$$
$$\left. + 242K_6 - 27K_7)\right)$$

$$K_9 = F\left(x_i + \frac{5}{6}h, y_i + \frac{1}{324}h(293K_0 - 852K_3 - 1375K_4 + 1836K_5 - 118K_6 \right.$$
$$\left. + 162K_7 + 324K_8)\right)$$

$$K_{10} = F\left(x_i + \frac{5}{6}h, y_i + \frac{1}{1620}h(1303K_0 - 4260K_3 - 6875K_4 + 9990K_5 + 1030K_6 \right.$$
$$\left. + 162K_9)\right)$$

$$K_{11} = F\left(x_i + h, y_i + \frac{1}{4428}h(-8595K_0 + 30720K_3 + 48\,750K_4 - 66\,096K_5 \right.$$
$$\left. + 378K_6 - 729K_7 - 1944K_8 - 1296K_9 + 3240K_{10})\right).$$

Formulas of order greater than 8 with exact coefficients are not readily available. These are given more conveniently with approximate coefficients (Curtis (1975), Hairer (1978)). For this reason we now give the Curtis formula of order 10 in the following fashion. To avoid notational confusion, the formula is given for the first step only.

$$y_1 = y_0 + h \sum_{i=1}^{18} b_i K_i$$

2 Runge–Kutta Methods

in which

$$K_1 = F(x_0, y_0)$$

$$K_i = F\left(x_0 + c_i h, y_0 + h \sum_{j=1}^{i-1} a_{ij} K_j\right), \quad i = 2, 3, 4, \ldots, 18$$

and the coefficients a_{ij}, b_i, c_i are given in Table 2.2.

Table 2.2 Coefficients for tenth-order explicit process.

$a_{2,1}$	= +0.14525 18960 31615 05176		$a_{9,7}$	= +0.39510 98495 81567 45999
c_2	= +0.14525 18960 31615 05176		$a_{9,8}$	= −0.04079 41270 37085 63576
$a_{3,1}$	= +0.07262 59480 15807 52588		c_9	= +0.58018 31400 82995 70863
$a_{3,2}$	= +0.07262 59480 15807 52588		$a_{10,1}$	= +0.07233 14442 23379 48078
c_3	= +0.14525 18960 31615 05176		$a_{10,2}$	= 0
$a_{4,1}$	= +0.05446 94610 11855 64441		$a_{10,3}$	= 0
$a_{4,2}$	= 0		$a_{10,4}$	= 0
$a_{4,3}$	= +0.16340 83830 35566 93323		$a_{10,5}$	= 0
c_4	= +0.21787 78440 47422 57764		$a_{10,6}$	= +0.22002 76284 68999 81021
$a_{5,1}$	= +0.54469 46101 18556 44411		$a_{10,7}$	= +0.08789 53342 54367 34013
$a_{5,2}$	= 0		$a_{10,8}$	= −0.04445 38399 62603 50864
$a_{5,3}$	= −2.04260 47879 44586 66540		$a_{10,9}$	= −0.21832 82289 48875 46891
$a_{5,4}$	= +2.04260 47879 44586 66540		c_{10}	= +0.11747 23380 35267 65357
c_5	= +0.54469 46101 18556 44411		$a_{11,1}$	= +0.08947 10093 67311 14229
$a_{6,1}$	= +0.06536 33532 14226 77329		$a_{11,2}$	= 0
$a_{6,2}$	= 0		$a_{11,3}$	= 0
$a_{6,3}$	= 0		$a_{11,4}$	= 0
$a_{6,4}$	= +0.32681 67660 71133 86646		$a_{11,5}$	= 0
$a_{6,3}$	= +0.26145 34128 56907 09317		$a_{11,6}$	= +0.39460 08170 28556 18607
c_6	= +0.65363 35321 42267 73293		$a_{11,7}$	= +0.34430 11367 96333 34877
$a_{7,1}$	= +0.08233 70775 74827 16585		$a_{11,8}$	= −0.07946 68266 42926 61291
$a_{7,2}$	= 0		$a_{11,9}$	= −0.39152 18947 89596 61238
$a_{7,3}$	= 0		$a_{11,10}$	= 0
$a_{7,4}$	= +0.21191 71963 20280 35617		c_{11}	= +0.35738 42417 59677 45184
$a_{7,5}$	= −0.03997 34350 80542 18312		$a_{12,1}$	= +0.03210 00687 79632 09213
$a_{7,6}$	= +0.02037 86531 75960 06198		$a_{12,2}$	= 0
c_7	= +0.27465 94919 90525 40088		$a_{12,3}$	= 0
$a_{8,1}$	= +0.08595 30577 90073 43832		$a_{12,4}$	= 0
$a_{8,2}$	= 0		$a_{12,5}$	= 0
$a_{8,3}$	= 0		$a_{12,6}$	= 0
$a_{8,4}$	= 0		$a_{12,7}$	= 0
$a_{8,5}$	= 0		$a_{12,8}$	= −0.00018 46375 99751 20501
$a_{8,6}$	= +0.29117 69478 05885 09603		$a_{12,9}$	= +0.15608 94025 31321 98608
$a_{8,7}$	= +0.39644 75145 14702 41049		$a_{12,10}$	= +0.19344 96857 65456 02528
c_8	= +0.77357 75201 10660 94484		$a_{12,11}$	= +0.26116 12387 63663 64969
$a_{9,1}$	= +0.08612 09348 56069 67550		c_{12}	= +0.64261 57582 40322 54816
$a_{9,2}$	= 0		$a_{13,1}$	= +0.04423 74932 85249 96327
$a_{9,3}$	= 0		$a_{13,2}$	= 0
$a_{9,4}$	= 0		$a_{13,3}$	= 0
$a_{9,5}$	= 0		$a_{13,4}$	= 0
$a_{9,6}$	= +0.13974 64826 82444 20890		$a_{13,5}$	= 0

2.3 Higher-Order Runge–Kutta Formulas

$a_{13,6} = 0$
$a_{13,7} = 0$
$a_{13,8} = +0.00464\,07744\,34539\,03964$
$a_{13,9} = +0.04704\,66028\,26151\,36532$
$a_{13,10} = +0.08620\,74994\,80114\,88160$
$a_{13,11} = -0.02607\,98302\,46821\,38093$
$a_{13,12} = -0.03858\,02017\,43966\,21532$
$c_{13} = +0.11747\,23380\,35267\,65357$
$a_{14,1} = +0.02318\,04671\,74294\,11567$
$a_{14,2} = 0$
$a_{14,3} = 0$
$a_{14,4} = 0$
$a_{14,5} = 0$
$a_{14,6} = 0$
$a_{14,7} = 0$
$a_{14,8} = +0.31978\,56784\,11636\,70673$
$a_{14,9} = +0.59332\,33331\,84189\,86861$
$a_{14,10} = -0.11171\,27832\,23452\,87347$
$a_{14,11} = +0.18039\,50557\,03050\,23573$
$a_{14,12} = -0.45540\,14298\,85722\,07269$
$a_{14,13} = +0.33295\,73406\,00736\,36584$
$c_{14} = +0.88252\,76619\,64732\,34643$
$a_{15,1} = +0.02624\,36432\,57981\,05892$
$a_{15,2} = 0$
$a_{15,3} = 0$
$a_{15,4} = 0$
$a_{15,5} = 0$
$a_{15,6} = 0$
$a_{15,7} = 0$
$a_{15,8} = +0.04863\,13942\,38672\,66107$
$a_{15,9} = +0.04274\,38253\,83464\,78868$
$a_{15,10} = -0.15575\,14733\,74349\,19785$
$a_{15,11} = +0.13260\,47194\,91765\,23318$
$a_{15,12} = -0.09402\,96215\,29465\,15652$
$a_{15,13} = +0.36891\,19544\,21896\,67414$
$a_{15,14} = -0.01197\,02001\,30288\,60976$
$c_{15} = +0.35738\,42417\,59677\,45184$
$a_{16,1} = +0.10284\,18616\,86822\,30957$
$a_{16,2} = 0$
$a_{16,3} = 0$
$a_{16,4} = 0$
$a_{16,5} = 0$
$a_{16,6} = 0$
$a_{16,7} = 0$
$a_{16,8} = -1.29708\,67206\,53005\,11984$
$a_{16,9} = -3.30609\,69033\,14255\,58655$
$a_{16,10} = -0.06747\,08496\,92334\,33385$
$a_{16,11} = -3.59679\,15480\,63726\,08732$
$a_{16,12} = +4.00369\,56199\,26740\,87775$
$a_{16,13} = +0.04100\,57488\,26127\,81166$
$a_{16,14} = +0.26978\,32108\,90450\,67539$
$a_{16,15} = +4.49273\,53386\,33502\,00133$
$c_{16} = +6.64261\,57582\,40322\,54816$

$a_{17,1} = +0.00876\,60933\,97736\,46361$
$a_{17,2} = 0$
$a_{17,3} = 0$
$a_{17,4} = 0$
$a_{17,5} = 0$
$a_{17,6} = 0$
$a_{17,7} = 0$
$a_{17,8} = +0.58404\,29244\,96019\,59632$
$a_{17,9} = +1.29877\,96899\,66251\,57393$
$a_{17,10} = -0.05854\,75188\,23066\,37183$
$a_{17,11} = +0.96649\,29364\,08446\,17558$
$a_{17,12} = -1.41862\,48359\,70972\,65394$
$a_{17,13} = +0.32460\,19941\,17747\,10444$
$a_{17,14} = -0.05497\,11264\,10616\,89342$
$a_{17,15} = -0.91544\,14072\,48110\,49093$
$a_{17,16} = +0.14742\,89120\,31297\,84267$
$c_{17} = +0.88252\,76619\,64732\,34643$
$a_{18,1} = +0.10173\,66974\,11157\,66388$
$a_{18,2} = 0$
$a_{18,3} = 0$
$a_{18,4} = 0$
$a_{18,5} = 0$
$a_{18,6} = 0$
$a_{18,7} = 0$
$a_{18,8} = -1.69621\,75532\,09432\,81071$
$a_{18,9} = -3.82523\,58462\,11624\,25452$
$a_{18,10} = -0.05854\,75188\,23066\,37183$
$a_{18,11} = -2.52076\,77892\,27152\,29120$
$a_{18,12} = +5.06895\,15710\,79828\,19146$
$a_{18,13} = +0.03221\,83852\,50196\,89969$
$a_{18,14} = +0.09661\,35792\,25014\,27296$
$a_{18,15} = +3.44722\,70365\,27756\,71816$
$a_{18,16} = -0.20173\,27870\,73975\,92774$
$a_{18,17} = +0.55575\,42250\,51297\,90987$
$c_{18} = +1.00000\,00000\,00000\,00000$
$b_1 = +0.03333\,33333\,33333\,33333$
$b_2 = 0$
$b_3 = 0$
$b_4 = 0$
$b_5 = 0$
$b_6 = 0$
$b_7 = 0$
$b_8 = 0$
$b_9 = 0$
$b_{10} = 0$
$b_{11} = 0$
$b_{12} = +0.23119\,09904\,31452\,64709$
$b_{13} = +0.18923\,74781\,48923\,49016$
$b_{14} = +0.03153\,95796\,91487\,24836$
$b_{15} = +0.27742\,91885\,17743\,17651$
$b_{16} = +0.04623\,81980\,86290\,52942$
$b_{17} = +0.15769\,78984\,57436\,24180$
$b_{18} = +0.03333\,33333\,33333\,33333$

For the above solution the arbitrary parameters were made definite by imposing the following conditions: $c_2 = c_3$; $c_5 = \frac{5}{6}c_6$; $b_{14} = \frac{1}{5}b_{17}$; $b_{16} = \frac{1}{5}b_{12}$; $a_{11,10} = 0$; $a_{18,10} = a_{17,10}$.

Other convenient formulas available in the literature have been developed by Fehlberg ((1966), (1968)), Hairer (1978), Huta (1957), Luther ((1966), (1968)), and others.

Of theoretical and practical interest is the following general result of Butcher (1987): *if one wishes to develop a Runge–Kutta formula of order n and if m is the number of K's one wishes to use, then $m > n$ whenever $n \geq 5$.*

2.4
Kutta's Fourth-Order Formula

Of all the Runge–Kutta formulas available, the one which seems to be most preferred, at least initially, is Kutta's fourth-order formula

$$y_{i+1} = y_i + \frac{1}{6}(K_0 + 2K_1 + 2K_2 + K_3) \tag{2.27}$$

in which

$$K_0 = hF(x_i, y_i)$$
$$K_1 = hF\left(x_i + \frac{1}{2}h, y_i + \frac{1}{2}K_0\right)$$
$$K_2 = hF\left(x_i + \frac{1}{2}h, y_i + \frac{1}{2}K_1\right)$$
$$K_3 = hF(x_{i+1}, y_i + K_2).$$

The combination of simplicity, relatively high accuracy, ease in programming, and economy in the use of (2.27) make it the usual first choice of many numerical analysts. For this reason, the considerations in the remainder of this chapter will focus on the use and generalizations of (2.27).

For illustrative purposes, let us apply (2.27) to the initial value problem

$$y' + y = x, \quad y(0) = 1.$$

Using $h = 0.1$ in $0 \leq x \leq 1$ and rounding to two decimal places, the formulas become

$$K_0 = (0.1)[-y_i + x_i]$$
$$K_1 = (0.1)\left[-y_i - \frac{1}{2}K_0 + x_i + 0.05\right]$$
$$K_2 = (0.1)\left[-y_i - \frac{1}{2}K_1 + x_i + 0.05\right]$$
$$K_3 = (0.1)[-y_i - K_2 + x_i + 0.1]$$
$$y_{i+1} = y_i + \frac{1}{6}[K_0 + 2K_1 + 2K_2 + K_3].$$

Computation for $i = 0, 1, 2, 3, 4, 5, 6, 7, 8, 9$ then yields $y_1 = 0.90967$, $y_2 = 0.83746$, $y_3 = 0.78164$, $y_4 = 0.74064$, $y_5 = 0.71306$, $y_6 = 0.69762$, $y_7 = 0.69317$, $y_8 = 0.69866$, $y_9 = 0.71314$, $y_{10} = 0.73576$, which agree with the exact solution $y = x - 1 + 2e^{-x}$ to five decimal places. In a rough way, this is consistent with the way in which the Kutta formula was constructed, for the numerical and analytical solutions agree through the h^4 terms in their Taylor expansions. Thus, the error behaves like terms of order h^5. In this example, $h = 0.1$ so that one would expect an error, at each calculation step, like $(0.1)^5 = 10^{-5}$. By changing the grid size to $h = 0.01$, one would expect an error at each time step of about $(0.01)^5 = 10^{-10}$.

2.5
Kutta's Formulas for Systems of First-Order Equations

For a system of two first-order equations with prescribed initial values, that is, for

$$y' = F(x, y, v), \quad y(0) = \alpha \quad (2.28)$$
$$v' = G(x, y, v), \quad v(0) = \beta, \quad (2.29)$$

the Kutta fourth-order technique can be extended directly. However, instead of employing extensive Taylor expansion considerations, which are required for a precise derivation, let us show how to do this easily by analogy. First, since there are two equations, we must have two sets of K's. For notational simplicity, we take, instead, a set of K's and a set of M's as follows:

$$K_0 = hF(x_i, y_i, v_i)$$
$$K_1 = hF\left(x_i + \frac{1}{2}h, y_i + \frac{1}{2}K_0, v_i + \frac{1}{2}M_0\right)$$
$$K_2 = hF\left(x_i + \frac{1}{2}h, y_i + \frac{1}{2}K_1, v_i + \frac{1}{2}M_1\right)$$
$$K_3 = hF(x_{i+1}, y_i + K_2, v_i + M_2)$$

$$M_0 = hG(x_i, y_i, v_i)$$
$$M_1 = hG\left(x_i + \frac{1}{2}h, y_i + \frac{1}{2}K_0, v_i + \frac{1}{2}M_0\right)$$
$$M_2 = hG\left(x_i + \frac{1}{2}h, y_i + \frac{1}{2}K_1, v_i + \frac{1}{2}M_1\right)$$
$$M_3 = hG(x_{i+1}, y_i + K_2, v_i + M_2).$$

Note that the formulas for the K's are related directly to (2.28) while those for the M's are related directly to (2.29), that is, the formulas for the K's use the function $F(x, y, v)$ in (2.28) while the formulas for the M's use the function $G(x, y, v)$ in (2.29). The calculation of y_{i+1} and v_{i+1} is given finally by

$$y_{i+1} = y_i + \frac{1}{6}(K_0 + 2K_1 + 2K_2 + K_3) \tag{2.30}$$

$$v_{i+1} = v_i + \frac{1}{6}(M_0 + 2M_1 + 2M_2 + M_3) \tag{2.31}$$

In the fashion described above, Kutta's formulas can be developed for arbitrarily large systems of first-order equations. For example for the system of three equations

$$y' = F(x, y, v, w), \quad y(0) = \alpha \tag{2.32}$$
$$v' = G(x, y, v, w), \quad v(0) = \beta, \tag{2.33}$$
$$w' = H(x, y, v, w), \quad w(0) = \gamma, \tag{2.34}$$

the formulas would be

$$K_0 = hF(x_i, y_i, v_i, w_i)$$
$$K_1 = hF\left(x_i + \frac{1}{2}h, y_i + \frac{1}{2}K_0, v_i + \frac{1}{2}M_0, w_i + \frac{1}{2}N_0\right)$$
$$K_2 = hF\left(x_i + \frac{1}{2}h, y_i + \frac{1}{2}K_1, v_i + \frac{1}{2}M_1, w_i + \frac{1}{2}N_1\right)$$
$$K_3 = hF(x_{i+1}, y_i + K_2, v_i + M_2, w_i + N_2)$$

$$M_0 = hG(x_i, y_i, v_i, w_i)$$
$$M_1 = hG\left(x_i + \frac{1}{2}h, y_i + \frac{1}{2}K_0, v_i + \frac{1}{2}M_0, w_i + \frac{1}{2}N_0\right)$$
$$M_2 = hG\left(x_i + \frac{1}{2}h, y_i + \frac{1}{2}K_1, v_i + \frac{1}{2}M_1, w_i + \frac{1}{2}N_1\right)$$
$$M_3 = hG(x_{i+1}, y_i + K_2, v_i + M_2, w_i + N_2)$$

$$N_0 = hH(x_i, y_i, v_i, w_i)$$
$$N_1 = hH\left(x_i + \frac{1}{2}h, y_i + \frac{1}{2}K_0, v_i + \frac{1}{2}M_0, w_i + \frac{1}{2}N_0\right)$$
$$N_2 = hH\left(x_i + \frac{1}{2}h, y_i + \frac{1}{2}K_1, v_i + \frac{1}{2}M_1, w_i + \frac{1}{2}N_1\right)$$
$$N_3 = hH(x_{i+1}, y_i + K_2, v_i + M_2, w_i + N_2)$$

and

$$y_{i+1} = y_i + \frac{1}{6}(K_0 + 2K_1 + 2K_2 + K_3)$$
$$v_{i+1} = v_i + \frac{1}{6}(M_0 + 2M_1 + 2M_2 + M_3)$$
$$w_{i+1} = w_i + \frac{1}{6}(N_0 + 2N_1 + 2N_2 + N_3).$$

A generic computer program for (2.30) and (2.31), which modifies directly for (2.27) is as follows:

Algorithm 2 Program Kutta Systems

Step 1. Set a counter $k = 1$.
Step 2. Set a time step h.
Step 3. Set an initial time x.
Step 4. Set initial values y, v.
Step 5. Calculate
$K_0 = hF(x, y, v)$
$M_0 = hG(x, y, v)$
$K_1 = hF\left(x + \frac{1}{2}h, y + \frac{1}{2}K_0, v + \frac{1}{2}M_0\right)$
$M_1 = hG\left(x + \frac{1}{2}h, y + \frac{1}{2}K_0, v + \frac{1}{2}M_0\right)$
$K_2 = hF\left(x + \frac{1}{2}h, y + \frac{1}{2}K_1, v + \frac{1}{2}M_1\right)$
$M_2 = hG\left(x + \frac{1}{2}h, y + \frac{1}{2}K_1, v + \frac{1}{2}M_1\right)$
$K_3 = hF(x + h, y + K_2, v + M_2)$
$M_3 = hG(x + h, y + K_2, v + M_2).$
Step 6. Calculate y at $x + h$ and v at $x + h$ by
$y(x+h) = y + \frac{1}{6}(K_0 + 2K_1 + 2K_2 + K_3)$
$v(x+h) = v + \frac{1}{6}(M_0 + 2M_1 + 2M_2 + M_3).$
Step 7. Increase the counter from k to $k+1$.
Step 8. Set $y = y(x+h), v = v(x+h), x = x+h$.
Step 9. Repeat Steps 5–8.
Step 10. Continue until $k = 100$.
Step 11. Stop the calculation.

2.6
Kutta's Formulas for Second-Order Differential Equations

Let us consider now the initial value problem

$$y'' = f(x, y, y') \tag{2.35}$$
$$y(0) = \alpha, \quad y'(0) = \beta. \tag{2.36}$$

If one first converts this problem to the equivalent system

$$y' = v, \quad y(0) = \alpha \tag{2.37}$$
$$v' = f(x, y, v), \quad v(0) = \beta \tag{2.38}$$

then Kutta's formulas for (2.28), (2.29) can be applied and, indeed, even simplified. The reason is that $F(x, y, v) = v$ in (2.37), and $F(x, y, v) = v$ is independent of x and y. Thus, for (2.37),

$$K_0 = h v_i$$
$$K_1 = h\left(v_i + \frac{1}{2}M_0\right)$$
$$K_2 = h\left(v_i + \frac{1}{2}M_1\right)$$
$$K_3 = h\left(v_i + M_2\right)$$

Substitution of these into (2.30) then implies that one need calculate only the M's. The following simple methodology then results.

To solve (2.35), (2.36) numerically, apply the Kutta formulas

$$y_{i+1} = y_i + h v_i + \frac{1}{6}h(M_0 + M_1 + M_2) \tag{2.39}$$

$$v_{i+1} = v_i + \frac{1}{6}(M_0 + 2M_1 + 2M_2 + M_3) \tag{2.40}$$

in which $v_i = y'_i$ and

$$M_0 = hf(x_i, y_i, v_i)$$
$$M_1 = hf\left(x_i + \frac{1}{2}h, y_i + \frac{1}{2}hv_i, v_i + \frac{1}{2}M_0\right)$$
$$M_2 = hf\left(x_i + \frac{1}{2}h, y_i + \frac{1}{2}hv_i + \frac{1}{4}hM_0, v_i + \frac{1}{2}M_1\right)$$
$$M_3 = hf\left(x_{i+1}, y_i + hv_i + \frac{1}{2}hM_1, v_i + M_2\right)$$

Example 2.2 With $h = 0.1$, let us show how to use Kutta's fourth-order formulas (2.39), (2.40) to approximate y and y' on $0 \leq x \leq 1$ for the initial value problem

$$y'' + xy' + y = 3 + 5x^2 \qquad (2.41)$$
$$y(0) = 1, \quad y'(0) = 0. \qquad (2.42)$$

Converting (2.41), (2.42) to a system yields

$$y' = v, \qquad\qquad y(0) = 1$$
$$v' = 3 + 5x^2 - xv - y, \quad v(0) = 0.$$

Thus,

$$f(x, y, v) = 3 + 5x^2 - xv - y.$$

Since $y_0 = 1$ and $v_0 = 0$, let $i = 0$ and consider $x_1 = 0.1$. Then

$M_0 = (0.1)f(x_0, y_0, v_0) = (0.1)f(0, 1, 0) = 0.2000$
$M_1 = (0.1)f(x_0 + 0.05, y_0 + 0.05v_0, v_0 + 0.1) = (0.1)f(0.05, 1, 0.1) = 0.2008$
$M_2 = (0.1)f(x_0 + 0.05, y_0 + 0.05v_0 + 0.025M_0, v_0 + 0.5M_1)$
$\quad = (0.1)f(0.05, 1.005, 0.1004) = 0.2002$
$M_3 = (0.1)f(x_1, y_0 + 0.1v_0 + 0.05M_1, v_0 + M_2)$
$\quad = (0.1)f(0.1, 1.01, 0.2002) = 0.2020,$

and, from (2.39), (2.40), one has to four decimal places,

$$y_1 = y_0 + (0.1)v_0 + \frac{1}{6}(0.1)(M_0 + M_1 + M_2) = 1.0100$$
$$v_1 = v_0 + \frac{1}{6}(M_0 + 2M_1 + 2M_2 + M_3) = 0.2007.$$

We next set $i = 1$ and proceed to the grid point $x_2 = 0.2$. The resulting new set of M's is

$M_0 = (0.1)f(x_1, y_1, v_1) = 0.2020$
$M_1 = (0.1)f(x_1 + 0.05, y_1 + 0.05v_1, v_1 + 0.5M_0) = 0.2047$
$M_2 = (0.1)f(x_1 + 0.05, y_1 + 0.05v_1 + 0.025M_0, v_1 + 0.5M_1) = 0.2042$
$M_3 = (0.1)f(x_2, y_1 + 0.1v_1 + 0.05M_1, v_1 + M_2) = 0.2079$

which yield

$$y_2 = 1.0403, \ v_2 = 0.4053.$$

Continuing in this fashion yields

$$y_3 = 1.0913, \quad v_3 = 0.6176$$
$$y_4 = 1.1642, \quad v_4 = 0.8410$$
$$y_5 = 1.2600, \quad v_5 = 1.0783$$
$$y_6 = 1.3804, \quad v_6 = 1.3318$$
$$y_7 = 1.5270, \quad v_7 = 1.6028$$
$$y_8 = 1.7015, \quad v_8 = 1.8921$$
$$y_9 = 1.9060, \quad v_9 = 2.1996$$
$$y_{10} = 2.1420, \quad v_{10} = 2.5246.$$

Algorithm 3 Kutta for Second-Order Equations

Step 1. Set a counter $k = 1$.
Step 2. Set a time step h.
Step 3. Set an initial time x.
Step 4. Set initial values y, v.
Step 5. Calculate
$$M_0 = hf(x, y, v)$$
$$M_1 = hf\left(x + \tfrac{1}{2}h, y + \tfrac{1}{2}hv, v + \tfrac{1}{2}M_0\right)$$
$$M_2 = hf\left(x + \tfrac{1}{2}h, y + \tfrac{1}{2}hv + \tfrac{1}{4}hM_0, v + \tfrac{1}{2}M_1\right)$$
$$M_3 = hf\left(x + h, y + hv + \tfrac{1}{2}hM_1, v + M_2\right).$$
Step 6. Calculate y at $x + h$ and v at $x + h$ by
$$y(x + h) = y + hv + \tfrac{1}{6}h(M_0 + M_1 + M_2)$$
$$v(x + h) = v + \tfrac{1}{6}(M_0 + 2M_1 + 2M_2 + M_3).$$
Step 7. Increase the counter from k to $k + 1$.
Step 8. Set $y = y(x + h)$, $v = v(x + h)$, $x = x + h$.
Step 9. Repeat Steps 5–8.
Step 10. Continue until $k = 100$.
Step 11. Stop the calculation.

2.7
Application – The Nonlinear Pendulum

Let us show now how to apply Kutta's formula to approximating the motion of a damped, nonlinear pendulum.

Consider a pendulum, as shown in Figure 2.1, which has mass m centered at P and is hinged at O. Assume that P is constrained to move on a circle of radius l whose center is O. Let θ be the angular measure, in radians, of the

2.7 Application – The Nonlinear Pendulum

pendulum's deviation from the vertical. The problem is that of describing the motion of P after release from a position of rest.

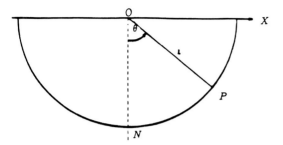

Fig. 2.1 Pendulum model.

It is known from laboratory experiments that the motion of the pendulum is damped and that the length of time between successive swings decreases.

Using cgs units, we reason analytically as follows. Assume that the motion of P is determined by Newton's dynamical equation

$$F = ma. \qquad (2.43)$$

Circular arc NP has length $l\theta$, so that $a = \frac{d^2}{dt^2}(l\theta) = l\ddot{\theta}$. Thus, (4.23) becomes

$$F = ml\ddot{\theta}. \qquad (2.44)$$

In considering the force F which acts on P, let F_1 be the gravitational component, so that

$$F_1 = -mg \sin\theta, \quad g > 0, \qquad (2.45)$$

and let F_2 be a damping component of the form

$$F_2 = -c\dot{\theta}, \quad c \text{ a nonnegative constant.} \qquad (2.46)$$

Assume that these are the only forces whose effects are significant. Then

$$F = -mg \sin\theta - c\dot{\theta},$$

so that (2.44) reduces readily to

$$\ddot{\theta} + \frac{c}{ml}\dot{\theta} + \frac{g}{l}\sin\theta = 0. \qquad (2.47)$$

The problem, then, is one of solving (2.47) subject to given initial conditions

$$\theta(0) = \alpha, \quad \dot{\theta}(0) = 0. \qquad (2.48)$$

For illustrative purposes, let us consider the strongly damped pendulum motion defined by

$$\ddot{\theta} + (0.3)\dot{\theta} + \sin\theta = 0 \qquad (2.49)$$

$$\theta(0) = \frac{1}{4}\pi, \quad \dot{\theta}(0) = 0. \qquad (2.50)$$

No analytical method is known for constructing the exact solution of this problem. Numerically, then, set $t = x$ and $\theta = y$ and solve (2.49) by the method of Section 2.6 with $h = 0.01$. The computation is carried out for 15000 time steps, that is, for 150 seconds of pendulum motion. The first 15.0 seconds of pendulum oscillation is shown in Figure 2.2, where the peak, or extreme, values $0.785\,40$, $-0.476\,47$, $0.293\,35$, $-0.181\,56$, $0.112\,59$ occur at the times $0.00, 3.28, 6.49, 9.68, 12.86$, respectively. The time required for the pendulum to travel from one peak to another decreases monotonically and damping is present during the entire simulation.

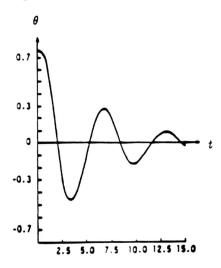

Fig. 2.2 Damped pendulum motion.

Any attempt to linearize (2.49) results in an analytical solution which either does not damp out, or has a constant time interval between successive swings, or both.

Nonlinear equations in all branches of science and engineering were often linearized in the days before computers, since the only general mathematical methodology for generating solutions was for linear equations. In this sense, the physics or chemistry of a model was distorted so that the mathematical equations could be solved. For example, in the areas of liquid and gas dynamics, the nonlinear Navier–Stokes equations were considered to be fundamental. Linearization resulted in equations for the flow of water which John von Neumann called the "equations of dry water". Further, insult was added to injury when these linearized equations were called "equations of ideal fluid flow", when there was nothing "ideal" about them at all.

2.8
Application – Impulsive Forces

A typical impulsive force problem of engineering interest is one in which there exists a simple discontinuity. For illustrative purposes, consider the particular initial value problem:

$$y'' + 2y' + 5y = \begin{cases} 1, & 0 < x < \pi \\ 0, & \pi < x < \infty \end{cases} \tag{2.51}$$

$$y(0) = y'(0) = 0. \tag{2.52}$$

In order to apply Kutta's formulas, all one need do is replace the differential equation (2.51) by

$$y'' + 2y' + 5y = \begin{cases} 1, & 0 < x < \pi \\ \frac{1}{2}, & x = \pi \\ 0, & \pi < x < \infty \end{cases} \tag{2.53}$$

The rationale for this is as follows. Physically, the right hand side of (2.53) is a force which is being applied to produce motion or change in position. Now this force is not in fact applied as given in (2.51) because an impulsive force is one whose values change continuously, but very quickly, over a short time period. Thus, the graph of the force is shown in Figure 2.3, whereas a more realistic graph would be that shown in Figure 2.4. For an impulsive force problem, the value of ϵ in Figure 2.4 would be exceptionally small. Nevertheless, the graph is continuous and smooth for all positive values of ϵ and if $\epsilon < \frac{1}{2}h$, then Kutta's formulas can be applied to (2.53) just as if the force were as shown in Figure 2.4.

Let us then apply Kutta's formulas (2.39), (2.40) with $h = (.001)\pi$. Such a choice requires a good approximation for π, which we take to be $\pi = 3.141\,592\,653\,589\,79$. The y and y' results are given each 100 time steps in columns B and D of Table 2.3 through x_{1500}. The Laplace transform result for problem (2.51), (2.52) is (Ross 1984, p. 450):

$$y = \begin{cases} \frac{1}{5}\left[1 - e^{-x}\left(\cos 2x + \frac{1}{2}\sin 2x\right)\right], & 0 < x < \pi \\ \frac{1}{5}e^{-x}\left[(e^\pi - 1)\cos 2x + \frac{1}{2}(e^\pi - 1)\sin 2x\right], & \pi < x < \infty. \end{cases} \tag{2.54}$$

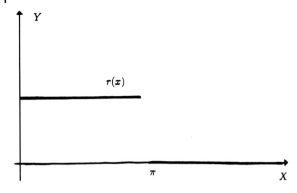

Fig. 2.3 Impulsive force.

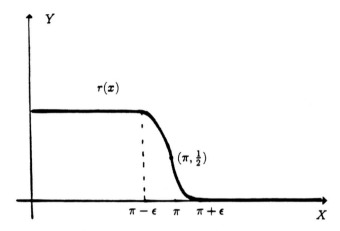

Fig. 2.4 Smoothed impulsive force.

The results for y and y' determined from (2.54) at the time steps in Table 2.3 are given in columns C and E of the table. Comparison of the numerical and transform results shows an exceptionally good agreement.

Table 2.3 Application of Kutta's formulas (2.39) and (2.40).

A Time	B y(Kutta)	C y(Trans.)	D y'(Kutta)	E y'(Trans.)
0.00000	0.00000	0.00000	0.00000	0.00000
0.31416	0.03889	0.03889	0.21466	0.21466
0.62832	0.11629	0.11629	0.25369	0.25369
0.94248	0.18702	0.18702	0.18529	0.18529
1.25664	0.22932	0.22932	0.08364	0.08364
1.57080	0.24158	0.24158	0.00000	0.00000
1.88496	0.23349	0.23349	−0.04462	−0.04462
2.19911	0.21740	0.21740	−0.05274	−0.05274
2.51327	0.20270	0.20270	−0.03852	−0.03852
2.82743	0.19390	0.19390	−0.01739	−0.01739
3.14159	0.19136	0.19136	0.00000	0.00000
3.14473	0.19136	0.19136	−0.00013	0.00014
3.14788	0.19135	0.19136	−0.00286	0.00000
3.45575	0.15482	0.15512	−0.20419	−0.20538
3.76991	0.08089	0.08060	−0.24300	−0.24273
4.08407	0.01300	0.01249	−0.17825	−0.17729
4.39823	−0.02779	−0.02823	−0.08102	−0.08003
4.71239	−0.03978	−0.04003	−0.00066	0.00000
5.02655	−0.03218	−0.03225	0.04245	0.04270
5.34071	−0.01682	−0.01675	0.05051	0.05046
5.65487	−0.00270	−0.00260	0.03705	0.03685
5.96903	0.00578	0.00587	0.01684	0.01664
6.28319	0.00827	0.00832	0.00014	0.00000
6.59734	0.00669	0.00670	−0.00882	−0.00888
6.91150	0.00350	0.00348	−0.01050	−0.01049
7.22566	0.00056	0.00054	−0.00770	−0.00766
7.53982	−0.00120	−0.00122	−0.00350	−0.00346
7.85398	−0.00172	−0.00173	−0.00003	0.00000
8.16814	−0.00139	−0.00139	0.00183	0.00185
8.48230	−0.00073	−0.00072	0.00218	0.00218
8.79646	−0.00012	−0.00011	0.00160	0.00159
9.11062	0.00025	0.00025	0.00073	0.00072
9.42478	0.00036	0.00036	0.00001	0.00000
9.73894	0.00029	0.00029	−0.00038	−0.00038
10.05310	0.00015	0.00015	−0.00045	−0.00045
10.36726	0.00002	0.00002	−0.00033	−0.00033
10.68142	−0.00005	−0.00005	−0.00015	−0.00015
10.99557	−0.00007	−0.00007	0.00000	0.00000
11.30973	−0.00006	−0.00006	0.00008	0.00008
11.62389	−0.00003	−0.00003	0.00009	0.00009
11.93805	−0.00001	0.00000	0.00007	0.00007
12.25221	0.00001	0.00001	0.00003	0.00003
12.56637	0.00002	0.00002	0.00000	0.00000
12.88053	0.00001	0.00001	−0.00002	−0.00002
13.19469	0.00001	0.00001	−0.00002	−0.00002
13.50885	0.00000	0.00000	−0.00001	−0.00001
13.82301	0.00000	0.00000	−0.00001	−0.00001
14.13717	0.00000	0.00000	0.00000	0.00000
14.45133	0.00000	0.00000	0.00000	0.00000
14.76549	0.00000	0.00000	0.00000	0.00000

2.9
Exercises

2.1 Using Kutta's formula with $h = 0.1$, find the numerical solution on $0 \le x \le 1$ for
$$y' = y^2 + 2x - x^4, \quad y(0) = 0.$$
Compare your results with those of Exercise 1 in Chapter 1.

2.2 Using Kutta's formula with $h = 0.05$, find the numerical solution on $0 \le x \le 1$ for
$$y' = xy^2 - 2y, \quad y(0) = 1.$$
Compare your results with those of Exercise 3 of Chapter 1.

2.3 Using Kutta's formula with $h = 0.01$, find the numerical solution on $0 \le x \le 10$ for
$$y' = y - \frac{2+x}{(1+x)^2}, \quad y(0) = 1.$$
Compare your results with those of Exercise 6 of Chapter 1.

2.4 Using Sarafyan's fifth-order formula (2.23) with $h = 0.05$, find the numerical solution on $0 \le x \le 3$ for
$$y' = y^2 - y - e^{-2x}, \quad y(0) = 1.$$
Compare your results with the exact solution $y = e^{-x}$ and verify that the solution is a fourth-order approximation at four interior grid points.

2.5 Using Butcher's sixth-order formula (2.24) with $h = 0.05$, find the numerical solution of the initial value problem in Exercise 4. Compare your results with the exact solution and with the solution of Sarafyan.

2.6 Using the eighth-order formula (2.26) of Shanks with $h = 0.05$, find the numerical solution of the initial value problem in Exercise 4. Compare your results with the exact solution $y = e^{-x}$. Discuss any problems which are present due to programming complexity and numerical rounding.

2.7 Derive a second-order Runge–Kutta formula which is different from (2.10). Apply your formula to the initial value problem in Exercise 1 and compare your results with those obtained by use of (2.10).

2.8 Using Kutta's formulas (2.30) and (2.31) for systems, find the numerical solution on $0 \le x \le 3$ with $h = 0.05$ of

$$y' = 1 + v - y^2 - v^2, \quad y(0) = 0$$
$$v' = 1 - y - y^2 - v^2, \quad v(0) = 1.$$

Compare your results with the exact solution $y = \sin x$, $v = \cos x$.

2.9 Using Kutta's formulas with $h = 0.05$, find the numerical solution on $0 \le x \le 2$ of the system

$$y' = 1 - y^2 + v, \quad y(0) = 0$$
$$v' = 2x + y^3 - w, \quad v(0) = 0$$
$$w' = 3x^2 + v - y^2, \quad w(0) = 0.$$

Compare your results with the exact solution $y = x$, $v = x^2$, $w = x^3$.

2.10 Using Kutta's formulas with $h = 0.05$, find the numerical solution on $0 \le x \le 2$ of the system

$$y' = 2x + w^2 - z^2, \quad y(0) = 0$$
$$v' = y + z, \quad v(0) = 0$$
$$w' = 2x + vx^2 + z, \quad w(0) = 0$$
$$z' = -2x + vx^2 - w, \quad z(0) = 0.$$

Compare your results with the exact solution $y = x^2$, $v = 1$, $w = x^2$, $z = -x^2$.

2.11 Using Kutta's formulas (2.39), (2.40) with $h = 0.01$, find the numerical solution of each of the following on $0 \le x \le 1$ and compare your results with the exact solution.

(a) $y'' = 2 + 8xy - (y')^3$, $y(0) = y'(0) = 0$ (Exact: $y = x^2$)
(b) $y'' = (x^2 - 1)y$, $y(0) = 1$, $y'(0) = 0$ (Exact: $y = e^{-\frac{1}{2}x^2}$)
(c) $y'' = x^3 - y^3$, $y(0) = y'(0) = 0$ (Exact: $y = x$)
(d) $y'' = 6x + 9x^4 - (y')^2$, $y(0) = y'(0) = 0$ (Exact: $y = x^3$).

2.12 Using Kutta's formulas with $h = 0.01$, find the numerical solution on $0 \le x \le 10$ for each of the following. Graph each solution and compare the results.

(a) $y'' + (0.2)y' + \sin y = 0$, $y(0) = \dfrac{1}{4}\pi$, $y'(0) = 0$

(b) $y'' + (0.1)y' + \sin y = 0$, $y(0) = \dfrac{1}{4}\pi$, $y'(0) = 0$

(c) $y'' + (0.05)y' + \sin y = 0$, $y(0) = \dfrac{1}{4}\pi$, $y'(0) = 0$

(d) $y'' + (0.01)y' + \sin y = 0$, $y(0) = \dfrac{1}{4}\pi$, $y'(0) = 0$.

3
The Method of Taylor Expansions

3.1
Introduction

The method of Taylor expansion, also called the method of Taylor series, is both old and well studied. Before the availability of modern computers, it was considered to be cumbersome because it often requires extensive differentiation. However, with the advent of computer techniques for *symbol manipulation*, by means of which the computer does the differentiation, the method of Taylor expansions has returned to the class of highly valuable methods (Corliss and Chang (1982); Moore (1979); Rall (1969)). In addition, not only does the method provide one with the means to check one's computations when done by, say, Runge–Kutta formulas, but it also provides one with a method which can be made to be of any order of accuracy, whereas Runge–Kutta formulas, at present, are available only up to order ten. The need for methods of exceptionally high orders of accuracy frequently occurs in astronomy and in astronautics.

3.2
First-Order Problems

Consider again the first-order initial value problem

$$y' = F(x,y), \quad y(0) = \alpha. \tag{3.1}$$

The method of Taylor expansions for the solution of (3.1) proceeds as follows. Assume the solution $y(x)$ has the Taylor expansion

$$y(x+h) = y(x) + hy'(x) + \frac{1}{2}h^2 y''(x) + \ldots + \frac{1}{k!}h^k y^{(k)}(x) + R_k. \tag{3.2}$$

Numerical Solution of Ordinary Differential Equations for Classical, Relativistic and Nano Systems. Donald Greenspan
Copyright © 2006 WILEY-VCH Verlag GmbH & Co. KGaA, Weinheim
ISBN: 3-527-40610-7

3 The Method of Taylor Expansions

The Taylor expansion method simply drops the term R_k in (3.2) and uses the kth-order formula

$$y(x+h) = y(x) + hy'(x) + \frac{1}{2}h^2 y''(x) + \ldots + \frac{1}{k!}h^k y^{(k)}(x). \tag{3.3}$$

In grid point notation, (3.3) becomes

$$y_{i+1} = y_i + hy'_i + \frac{1}{2}h^2 y''_i + \ldots + \frac{1}{k!}h^k y_i^{(k)}. \tag{3.4}$$

Formula (3.4) is called a kth-order approximation because, as discussed in Chapter 2, it uses the Taylor expansion terms for the exact solution through the terms of order h^k.

Example 3.1 To illustrate its implementation, consider again (1.11), that is,

$$y' + y = x, \quad y(0) = 1 \tag{3.5}$$

and let us find a numerical solution on $0 \leq x \leq 1$ with $h = 0.1$. Let us utilize a fourth-order expansion, so that we can compare the results with those obtained in Section 2.4 using Kutta's formula.

Solution. The grid points are $x_0 = 0.0$, $x_1 = 0.1$, $x_2 = 0.2, \ldots, x_{10} = 1.0$. The Taylor formula is

$$y_{i+1} = y_i + (0.1)y'_i + \frac{1}{2}(0.1)^2 y''_i + \frac{1}{6}(0.1)^3 y'''_i + \frac{1}{24}(0.1)^4 y_i^{iv},$$
$$i = 0, 1, 2, \ldots, 9, \tag{3.6}$$

with

$$y_0 = 1. \tag{3.7}$$

Since the computer will round the coefficients in (3.6), let us do so immediately to yield

$$y_{i+1} = y_i + (0.1)y'_i + (0.005)y''_i + (0.000\,166\,7)y'''_i + (0.000\,004\,1)y_i^{iv}. \tag{3.8}$$

Consider first $i = 0$. Then (3.8) yields

$$y_1 = y_0 + (0.1)y'_0 + (0.005)y''_0 + (0.000\,166\,7)y'''_0 + (0.000\,004\,1)y_0^{iv}. \tag{3.9}$$

However, though y_0 is known, we are not given $y'_0, y''_0, y'''_0, y_0^{iv}$. These have to be determined from the differential equation by the chain rule as follows. Since

$$y' = -y + x$$

then

$$y'' = -y' + 1$$
$$y''' = -y''$$
$$y^{iv} = -y''',$$

which, at the grid points, yields

$$\begin{aligned} y'_i &= -y_i + x_i \\ y''_i &= -y'_i + 1 \\ y'''_i &= -y''_i \\ y^{iv}_i &= -y'''_i. \end{aligned} \quad (3.10)$$

For $i = 0$, we have from (3.10)

$$\begin{aligned} y'_0 &= -y_0 + x_0 = -1 \\ y''_0 &= -y'_0 + 1 = 2 \\ y'''_0 &= -y''_0 = -2 \\ y^{iv}_0 &= -y'''_0 = 2, \end{aligned}$$

which, upon substitution into (3.9), yields

$$y_1 = 1 - (0.1) + 2(0.005) - 2(0.000\,166\,7) + 2(0.000\,004\,1) = 0.909\,675. \quad (3.11)$$

Continuing in the indicated fashion to generate $y_2, y_3, \ldots, y_{10}$, one finds readily that the numerical solution agrees with the Kutta results uniformly through four decimal places, as is to be expected. In generic form, a typical computer program is given as follows.

Algorithm 4 Program Taylor 1

Step 1. Set a counter $K = 1$.
Step 2. Set a time step h.
Step 3. Set an initial time x.
Step 4. Set an initial value y.
Step 5. Determine the first four derivatives y_1, y_2, y_3, y_4.
Step 6. Determine y at $x + h$ from

$$y(x+h) = y + hy_1 + \frac{1}{2!}h^2 y_2 + \frac{1}{3!}h^3 y_3 + \frac{1}{4!}h^4 y_4.$$

Step 7. Increase the counter from K to $K + 1$.
Step 8. Set $y = y(x+h)$, $x = x + h$.
Step 9. Repeat Steps 5–8.
Step 10. Continue until $K = 20$.
Step 11. Stop the calculation.

3.3
Systems of First-Order Equations

The method developed in Section 3.2 extends directly to systems of equations. To illustrate the ideas, consider the system

$$y' = F(x, y, v), \quad y(0) = \alpha \tag{3.12}$$

$$v' = G(x, y, v), \quad v(0) = \beta. \tag{3.13}$$

Assuming $y(x)$ and $v(x)$ have appropriate Taylor expansions, the kth-order approximation formulas one uses are

$$y_{i+1} = y_i + hy'_i + \frac{1}{2}h^2 y''_i + \ldots + \frac{1}{k!}h^k y^{(k)}, \quad i = 0, 1, 2, \ldots, n-1 \tag{3.14}$$

$$v_{i+1} = v_i + hv'_i + \frac{1}{2}h^2 v''_i + \ldots + \frac{1}{k!}h^k v^{(k)}, \quad i = 0, 1, 2, \ldots, n-1. \tag{3.15}$$

The higher order derivatives of y and v, necessary for the implementation of (3.14), (3.15) are obtained by differentiation of (3.12), (3.13).

Example 3.2 Consider the initial value problem

$$\frac{dy}{dx} = y - v + 2x + 1, \quad y(0) = 0 \tag{3.16}$$

$$\frac{dv}{dx} = y - x^2 + 2x, \quad v(0) = 1. \tag{3.17}$$

This problem does not require numerical methodology because it can be solved exactly to yield the solution $y = x^2$, $v = x^2 + 1$. We will proceed numerically only for illustrative purposes.

On $0 \leq x \leq 1$, let $h = 0.1$ and assume fourth-order Taylor expansions for both y and v, so that

$$y_{i+1} = y_i + (0.1)y'_i + \tfrac{1}{2}(0.1)^2 y''_i + \tfrac{1}{6}(0.1)^3 y'''_i + \tfrac{1}{24}(0.1)^4 y^{iv}_i,$$
$$i = 0, 1, 2, \ldots, 9, \tag{3.18}$$

$$v_{i+1} = v_i + (0.1)v'_i + \tfrac{1}{2}(0.1)^2 v''_i + \tfrac{1}{6}(0.1)^3 v'''_i + \tfrac{1}{24}(0.1)^4 v^{iv}_i,$$
$$i = 0, 1, 2, \ldots, 9. \tag{3.19}$$

From (3.16) and (3.17), then

$$y'_i = y_i - v_i + 2x_i + 1, \qquad v'_i = y_i - x_i^2 + 2x_i$$
$$y''_i = y'_i - v'_i + 2, \qquad v''_i = y'_i - 2x_i + 2$$
$$y'''_i = y''_i - v''_i, \qquad v'''_i = y''_i - 2$$
$$y^{iv}_i = y'''_i - v'''_i, \qquad v^{iv}_i = y'''_i.$$

For $i = 0$, one has

$$y_0 = 0, \quad v_0 = 1$$
$$y_0' = 0, \quad v_0' = 0$$
$$y_0'' = 2, \quad v_0'' = 2$$
$$y_0''' = 0, \quad v_0''' = 0$$
$$y_0^{iv} = 0, \quad v_0^{iv} = 0,$$

so that (3.18), (3.19) yield

$$y_1 = 0.0100, \quad v_1 = 1.0100.$$

One now continues, in the indicated fashion, to generate y_2, v_2, y_3, $v_3, \ldots, y_{10}, v_{10}$, which agree with the exact solution to four decimal places.

In generic form, a typical computer program is as follows.

Algorithm 5 Program Taylor System

Step 1.   Set a counter $K = 1$.
Step 2.   Set a time step h.
Step 3.   Set an initial time x.
Step 4.   Set initial values y and v.
Step 5.   From the differential equations, determine the first five derivatives y_1, y_2, y_3, y_4, y_5 of y, and the first five derivatives v_1, v_2, v_3, v_4, v_5 of v.
Step 6.   Determine y at $x + h$ and v at $x + h$ from

$$y(x+h) = y + hy_1 + \frac{1}{2!}h^2 y_2 + \frac{1}{3!}h^3 y_3 + \frac{1}{4!}h^4 y_4 + \frac{1}{5!}h^5 y_5$$

$$v(x+h) = v + hv_1 + \frac{1}{2!}h^2 v_2 + \frac{1}{3!}h^3 v_3 + \frac{1}{4!}h^4 v_4 + \frac{1}{5!}h^5 v_5.$$

Step 7.   Increase the counter from K to $K + 1$.
Step 8.   Set $y = y(x+h)$, $v = v(x+h)$, $x = x + h$.
Step 9.   Repeat Steps 5–8.
Step 10.   Continue until $K = 10$.
Step 11.   Stop the calculation.

3.4
Second-Order Initial Value Problems

Given the second-order initial value problem

$$y'' = F(x, y, y'), \quad y(0) = \alpha, \quad y'(0) = \beta, \quad (3.20)$$

the Taylor expansion method allows one to solve (3.20) numerically without converting to a first-order system, as was required by Kutta's method. The

3 The Method of Taylor Expansions

reason is that if y is sufficiently differentiable, then one has both

$$y(x+h) = y(x) + hy'(x) + \frac{1}{2}h^2 y''(x) + \ldots + \frac{1}{k!}h^k y^{(k)}(x) + R_1 \qquad (3.21)$$

$$y'(x+h) = y'(x) + hy''(x) + \frac{1}{2}h^2 y'''(x) + \ldots + \frac{1}{k!}h^k y^{(k+1)}(x) + R_2. \qquad (3.22)$$

The resulting kth-order numerical formulas, which follow by dropping the remainder terms in (3.21) and (3.22), are

$$y_{i+1} = y_i + hy'_i + \frac{1}{2}h^2 y''_i + \ldots + \frac{1}{k!}h^k y^{(k)} \qquad (3.23)$$

$$y'_{i+1} = y'_i + hy''_i + \frac{1}{2}h^2 y'''_i + \ldots + \frac{1}{k!}h^k y^{(k+1)}. \qquad (3.24)$$

The higher order derivatives in (3.23), (3.24) are found by differentiating the given differential equation.

Example 3.3 Consider again, on $0 \leq x \leq 1$, the initial value problem (2.41), (2.42), that is

$$y'' + xy' + y = 3 + 5x^2 \qquad (3.25)$$

$$y(0) = 1, \quad y'(0) = 0. \qquad (3.26)$$

Let us consider fourth-order Taylor expansions with $h = 0.1$, so that the particular formulas are

$$y_{i+1} = y_i + (0.1)y'_i + \frac{1}{2}(0.1)^2 y''_i + \frac{1}{6}(0.1)^3 y'''_i + \frac{1}{24}(0.1)^4 y^{iv}_i \qquad (3.27)$$

$$y'_{i+1} = y'_i + (0.1)y''_i + \frac{1}{2}(0.1)^2 y'''_i + \frac{1}{6}(0.1)^3 y^{iv}_i + \frac{1}{24}(0.1)^4 y^{v}_i \qquad (3.28)$$

The higher order derivatives in (3.27), (3.28) are found, by differentiation, to be,

$$y''_i = 3 + 5x_i^2 - y_i - x_i y'_i$$
$$y'''_i = 10x_i - 2y'_i - x_i y''_i$$
$$y^{iv}_i = 10 - 3y''_i - x_i y'''_i$$
$$y^{v}_i = -4y'''_i - x_i y^{iv}_i.$$

Thus, for $i = 0$, one has

$$y_0 = 1$$
$$y'_0 = 0$$
$$y''_0 = 3 + 5x_0^2 - y_0 - x_0 y'_0 = 2$$
$$y'''_0 = 10x_0 - 2y'_0 - x_0 y''_0 = 0$$
$$y^{iv}_0 = 10 - 3y''_0 - x_0 y'''_0 = 4$$
$$y^{v}_0 = -4y'''_0 - x_0 y^{iv}_0 = 0,$$

which, upon substitution into (3.27), (3.28) yields, to six decimal places,

$$y_1 = 1.010017, \quad y'_1 = 0.200667. \tag{3.29}$$

If one continues in this fashion for $i = 1, 2, \ldots, 9$, one finds to six decimal places

$$\begin{aligned}
x_0 &= 0.0 & y_0 &= 1.000\,000 & y'_0 &= 0.000\,000 \\
x_1 &= 0.1 & y_1 &= 1.010\,017 & y'_1 &= 0.200\,667 \\
x_2 &= 0.2 & y_2 &= 1.040\,265 & y'_2 &= 0.405\,284 \\
x_3 &= 0.3 & y_3 &= 1.091\,331 & y'_3 &= 0.617\,605 \\
x_4 &= 0.4 & y_4 &= 1.164\,157 & y'_4 &= 0.841\,010 \\
x_5 &= 0.5 & y_5 &= 1.259\,999 & y'_5 &= 1.078\,341 \\
x_6 &= 0.6 & y_6 &= 1.380\,364 & y'_6 &= 1.331\,789 \\
x_7 &= 0.7 & y_7 &= 1.526\,945 & y'_7 &= 1.602\,814 \\
x_8 &= 0.8 & y_8 &= 1.701\,538 & y'_8 &= 1.892\,111 \\
x_9 &= 0.9 & y_9 &= 1.905\,976 & y'_9 &= 2.199\,630 \\
x_{10} &= 1.0 & y_{10} &= 2.142\,049 & y'_{10} &= 2.524\,626
\end{aligned}$$

These agree with those generated by Kutta's formulas in Section 2.6 to five decimal places.

3.5
Application – The van der Pol Oscillator

A van der Pol oscillator is a solution of the equation

$$y'' - \lambda(1 - y^2)y' + y = 0, \tag{3.30}$$

in which λ is a positive constant. This equation is called the van der Pol equation and is an equation which models a nonlinear diode oscillator. The problem of interest relative to (3.30) is to find a periodic solution for each given λ (Greenspan and Casulli (1988); Urabe (1960), (1961)). Relative to (3.30) the following result is known (Clenshaw (1966)). For each $\lambda > 0$, there exists exactly one periodic solution of (3.30). Moreover, if $y'(0) = 0$, $y(0) = \alpha > 0$, and $T = T(\lambda)$ is the period of $y(x)$, then $y'(\frac{1}{2}T) = 0$, $y(\frac{1}{2}T) = -\alpha$, and $y(x)$ is monotonic decreasing on $0 \leq x \leq \frac{1}{2}T$, while it is monotonic increasing on $\frac{1}{2}T \leq x \leq T$. The essential content of Clenshaw's result is shown in Figure 3.1 for the half period $0 \leq x \leq \frac{1}{2}T$ and motivates the following numerical approach to generating a periodic solution for given λ.

Set $y'(0) = 0$. Next, choose the following approximations for α: $\alpha_1 = 1, \alpha_2 = 2, \alpha_3 = 3, \ldots, \alpha_{10} = 10$. Beginning with the smallest α, that is with $\alpha_1 = 1$,

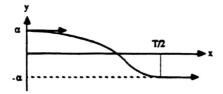

Fig. 3.1 Graphical representation of Clenshaw's theorem.

solve numerically, and in sequence, the set of eleven initial value problems,

$$y'' - \lambda(1-y^2)y' + y = 0, \quad y(0) = \alpha_i, \quad y'(0) = 0, \quad i = 1, 2, 3, \ldots, 10. \quad (3.31)$$

For each α_i, let β_i be the first minimum value which the numerical solution yields. Figure 3.2 shows possible results corresponding to α_1, α_2, α_3. Let $\frac{1}{2}T_i$ be the x value corresponding to β_i. In Figure 3.2 note that $\alpha_1 < -\beta_1$, $\alpha_2 < -\beta_2$, $\alpha_3 > -\beta_3$. We want $\alpha_i = \beta_i$. Hence the desired value of α must be in the range $2 \leq \alpha \leq 3$. We next consider a new set of α parameters, namely, $\alpha_0 = 2.0, \alpha_1 = 2.1, \alpha_2 = 2.2, \ldots, \alpha_{10} = 3.0$, and solve in sequence the initial value problems (3.31). For each α_i, let β_i be the first minimum value which the numerical solution yields. As above, comparison of each α_i with the corresponding $-\beta_i$ leads to new bounds on the exact α sought. For example, we may find that $\alpha_6 \leq \alpha \leq \alpha_7$, that is $2.6 \leq \alpha \leq 2.7$. We would then continue by considering $\alpha_0 = 2.60, \alpha_1 = 2.61, \alpha_2 = 2.62, \ldots, \alpha_{10} = 2.70$, and repeat the process in the indicated fashion. In this way we can find an approximation α_i to α to any fixed number of decimal places. The corresponding $x = \frac{1}{2}T_i$ at which $y(\frac{1}{2}T_i) = \beta_i$ yields the half period.

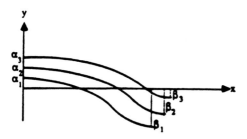

Fig. 3.2 The α and β correspondance.

Example 3.4 Consider the van der Pol equation for each of the three values $\lambda=0.1$, $\lambda=1$, $\lambda=10$. We wish to approximate α and $\frac{1}{2}T$ for each. In implementing the iterative method described above we apply fourth-order Taylor expansion

formulas with $h = 0.001$. For this purpose note that
$$y'' = \lambda y' - \lambda y^2 y' - y$$
$$y''' = \lambda y'' - 2\lambda y(y')^2 - \lambda y^2 y'' - y'$$
$$y^{iv} = \lambda y''' - 2\lambda(y')^3 - 6\lambda yy'y'' - \lambda(y)^2 y''' - y''$$
$$y^{v} = \lambda y^{iv} - 12\lambda (y')^2 y'' - 6\lambda y(y'')^2 - 8\lambda yy'y''' - \lambda y^2 y^{iv} - y'''.$$

The numerical results which follow are summarized in Table 3.1.

Table 3.1 Numerical results for the van der Pol equation for three values of λ.

λ	α	$\frac{1}{2}T$
0.1	2.000	3.148
1.0	2.009	3.335
10.0	2.014	9.538

The graphs of the approximate periodic solutions are shown in Figure 3.3. As a check, the same results are obtained using Kutta's formulas.

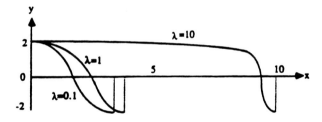

Fig. 3.3 Solutions of van der Pol's equation.

Of special interest are the results for $\lambda = 10$. These results reveal the property called stiffness, in which one of the variables is changing slowly relative to the other variable, which is changing relatively rapidly. Indeed, until approximately $x = 8$, y decreases slowly as x increases relatively rapidly, while in the range $8 < x < 10$, y changes rapidly relative to small changes in x.

3.6
Exercises

3.1 Using a fourth-order Taylor expansion with $h = 0.1$, find the numerical solution on $0 \le x \le 1$ for
$$y' = y^2 + 2x - x^4, \quad y(0) = 0.$$
Compare your results with those of Exercise 1 of Chapter 2.

3 The Method of Taylor Expansions

3.2 Using a fifth-order Taylor expansion with $h = 0.05$, find the numerical solution on $0 \leq x \leq 3$ for

$$y' = y^2 - y - e^{-2x}, \quad y(0) = 1.$$

Compare your results with the exact solution $y = e^{-x}$.

3.3 Using a sixth-order Taylor expansion with $h = 0.05$, find the numerical solution of the initial value problem in Exercise 2 and compare your results with those of Exercise 2 and with the exact solution.

3.4 Using an eighth-order Taylor expansion with $h = 0.05$, find the numerical solution of the initial value problem in Exercise 2 and compare your results with those of Exercise 2 and with the exact solution.

3.5 Using fourth-order Taylor expansions, find the numerical solution on $0 \leq x \leq 3$ with $h = 0.05$ of

$$y' = 1 + v - y^2 - v^2, \quad y(0) = 0$$
$$v' = 1 - y - y^2 - v^2, \quad v(0) = 1.$$

Compare your results with the exact solution $y = \sin x$, $v = \cos x$.

3.6 Using fourth-order Taylor expansions, find the numerical solution on $0 \leq x \leq 2$ with $h = 0.05$ of

$$y' = 1 - y^2 + v, \quad y(0) = 0$$
$$v' = 2x + y^3 - w, \quad v(0) = 0$$
$$w' = 3x^2 + v - y^2, \quad w(0) = 0.$$

Compare your results with the exact solution $y = x$, $v = x^2$, $w = x^3$.

3.7 Using fourth-order Taylor expansions with $h = 0.1$, find the numerical solution on $0 \leq x \leq 1$ for each of the following, and compare your results with the exact solution:

(a) $y'' = 2 + 8xy - (y')^3$, $y(0) = y'(0) = 0$, (Exact: $y = x^2$)
(b) $y'' = (x^2 - 1)y$, $y(0) = 1, y'(0) = 0$, (Exact: $y = e^{-\frac{1}{2}x^2}$)
(c) $y'' = x^3 - y^3$, $y(0) = y'(0) = 0$, (Exact: $y = x$)
(d) $y'' = 6x + 9x^4 - (y')^2$, $y(0) = y'(0) = 0$, (Exact: $y = x^3$)

3.8 Using fourth-order Taylor expansions with $h = 0.01$, find the numerical solution on $0 \leq x \leq 10$ for each of the following. Graph each and compare the results.

(a) $y'' + (0.2)y' + \sin y = 0$, $\quad y(0) = \dfrac{1}{4}\pi, \quad y'(0) = 0$

(b) $y'' + (0.05)y' + \sin y = 0$, $\quad y(0) = \dfrac{1}{4}\pi, \quad y'(0) = 0.$

4
Large Second-Order Systems with Application to Nano Systems

4.1
Introduction

The methodology to be developed in this chapter can be applied to any large system of second-order ordinary differential equations. We will develop it in conjunction with the currently important area of nano mechanics.

4.2
The N-Body Problem

The fundamental mathematical problem in the study of nano systems is a noncontinuum problem called the N-body problem, which is described in complete generality as follows. In cgs units and for $i = 1, 2, \ldots, N$, let P_i of mass m_i be at $\vec{r}_i = (x_i, y_i, z_i)$, have velocity $\vec{v}_i = (v_{i,x}, v_{i,y}, v_{i,z})$, and have acceleration $\vec{a}_i = (a_{i,x}, a_{i,y}, a_{i,z})$ at time $t \geq 0$. Let the positive distance between P_i and P_j, $i \neq j$, be $r_{ij} = r_{ji} \neq 0$. Let the force on P_i due to P_j be $\vec{F}_{ij} = \vec{F}_{ij}(r_{ij})$, so that the force depends only on the distance between P_i and P_j. Also, assume that the force $\vec{F}_{ji}$ on P_j due to P_i satisfies $\vec{F}_{ji} = -\vec{F}_{ij}$. Then, given the initial positions and velocities of all the P_i, $i = 1, 2, 3, \ldots, N$, the general N-body problem is to determine the motion of the system if each P_i interacts with all the other P_j's in the system.

A nano system is an N-body problem in which the P_i are atoms or molecules.

The prototype N-body problem was formulated in about 1900. In it the P_i were the sun and the then known eight planets, and the force on each P_i was gravitational attraction. This problem is exceptionally difficult for $N \geq 3$. The additional difficulties with the problems to be considered in this chapter arise from the fact that we will be concerned with the interactions of molecules, for which the forces are more complex than gravitation.

Numerical Solution of Ordinary Differential Equations for Classical, Relativistic and Nano Systems. Donald Greenspan
Copyright © 2006 WILEY-VCH Verlag GmbH & Co. KGaA, Weinheim
ISBN: 3-527-40610-7

4.3
Classical Molecular Potentials

Classical molecular forces behave, in general, as follows (Feynman et al. (1973)). When two close (to be made precise shortly) molecules are pulled apart, they attract. When pushed together, they repel. And the force of repulsion is of a greater order of magnitude than the force of attraction.

Fig. 4.1 Two molecules on an X axis.

Example 4.1 Consider two hypothetical molecules P_1, P_2 on an X axis, as shown in Figure 4.1. Let P_1 be at the origin and let P_2 be R units from P_1, $R > 0$. Let $\vec{F}$ be the force P_1 exerts on P_2 and let F be the magnitude of $\vec{F}$. Suppose

$$F = \frac{1}{R^{13}} - \frac{1}{R^7}. \tag{4.1}$$

Then, if $R = 1$, $F = 0$, and the molecules are in equilibrium. If $R > 1$, say, $R = 2$, then

$$F = \frac{1}{2^{13}} - \frac{1}{2^7} < 0,$$

so that $\vec{F}$ acts toward the origin, which corresponds to attraction. If $R < 1$, say, $R = 0.1$, then

$$F = \frac{1}{0.1^{13}} - \frac{1}{0.1^7} > 0,$$

so that $\vec{F}$ acts away from the origin, which corresponds to repulsion. Note that F is unbounded as R converges to zero, so that as R converges to zero the interaction of the two molecules can be extremely volatile.

There are a variety of classical molecular potentials for the interactions of molecules and from these classical molecular force formulas can be derived (Hirschfelder et al. (1967)). There are, for example, Buckingham, Lennard–Jones, Morse, Slater–Kirkwood, Stockmayer, Sutherland, and Yntema-Schneider potentials. The potential which has received the most attention is the Lennard–Jones (6,12) potential, that is,

$$\phi(r_{ij}) = 4\epsilon \left[\frac{\sigma^{12}}{r_{ij}^{12}} - \frac{\sigma^6}{r_{ij}^6} \right] \text{erg,}$$

some examples of which can be found in Table 4.1.

Let us turn our attention first to a Lennard–Jones potential for argon vapor. We examine this first because it is thought that for argon the Lennard–Jones potential is quantitatively accurate (Koplik and Banavar (1998)). The potential is

$$\phi(r_{ij}) = (6.848)10^{-14}\left[\frac{3.418^{12}}{r_{ij}^{12}} - \frac{3.418^6}{r_{ij}^6}\right] \text{erg} \quad \left(\frac{\text{g cm}^2}{\text{s}^2}\right) \quad (4.2)$$

in which r_{ij} is measured in angstroms (Å). (Recall that 1 cm = 10^8 Å.) The force $\vec{F}_{ij}$ exerted on P_i by P_j is then

$$\vec{F}_{ij} = (6.848)10^{-14}\left[\frac{12(3.418)^{12}}{r_{ij}^{13}} - \frac{6(3.418)^6}{r_{ij}^7}\right](10)^8 \frac{\vec{r}_{ji}}{r_{ij}} \text{ dynes} \quad \left(\frac{\text{g cm}}{\text{s}^2}\right). \quad (4.3)$$

Note that in deriving (4.3) from (4.2), one must use the chain rule

$$F_{ij} = -\frac{d\phi(r_{ij})}{dR} = -\frac{d\phi(r_{ij})}{dr_{ij}}\frac{dr_{ij}}{dR}$$

and the fact $r_{ij} = 10^8 R$. Hence, (4.3) reduces readily to

$$\vec{F}_{ij} = \left[\frac{209.0}{r_{ij}^{13}} - \frac{0.06551}{r_{ij}^7}\right]\frac{\vec{r}_{ji}}{r_{ij}} \text{ dynes} \quad \left(\frac{\text{g cm}}{\text{sec}^2}\right). \quad (4.4)$$

Note also that

$$F_{ij} = \|\vec{F}_{ij}\| = \left[\frac{209.0}{r_{ij}^{13}} - \frac{0.06551}{r_{ij}^7}\right],$$

so that $F_{ij}(r_{ij}) = 0$ implies that $r_{ij} = 3.837$ Å, which is the equilibrium distance.

Table 4.1 Lennard–Jones (6,12) potentials:
$\phi(r_{ij}) = 4\epsilon\left[\frac{\sigma^{12}}{r_{ij}^{12}} - \frac{\sigma^6}{r_{ij}^6}\right]$ erg, $k = (1.381)10^{-16}$.

Gas	ϵ/k (°K)	σ (Å)
Ar	124	3.418
Ne	35.7	2.789
CO	110	3.590
CO_2	190	3.996
NO	119	3.470
CH_4	137	3.822
SO_2	252	4.290
F_2	112	3.653
Cl_2	357	4.115
C_6H_6	440	5.270
Air	97.0	3.617

4.4
Molecular Mechanics

Molecular mechanics is the simulation of molecular interaction as an N-body problem using classical molecular potentials and Newtonian mechanics. Unlike quantum mechanics, molecular mechanics is deterministic. For clarity in molecular mechanics simulations, recall that for simple fluid flows, the temperature T on the Kelvin scale of a molecule of mass m g and speed v cm/s is given in two dimensions by

$$kT = \frac{1}{2}mv^2 \tag{4.5}$$

in which k is the Boltzmann constant $(1.381)10^{-16}$ erg deg^{-1}. Also recall that T kelvin and C degrees centigrade are related by

$$T = C + 273. \tag{4.6}$$

4.5
The Leap Frog Formulas

Classical molecular force formulas require small time steps in any numerical simulation in order to yield physically correct results for the effect of repulsion, which is unbounded when the distance between the molecules is close to zero. Because we are restricted physically to small time steps and because the number of equations is usually exceptionally large, Runge–Kutta and Taylor expansion methods, for example, prove to be unwieldy for related problems. Hence, we now describe the leap frog formulas which will be used throughout this chapter. Choose a positive time step h and let $t_k = kh$, $k = 0, 1, 2, \ldots$. For $i = 1, 2, 3, \ldots, N$, let P_i have mass m_i and at t_k let it be at $\vec{r}_{i,k} = (x_{i,k}, y_{i,k}, z_{i,k})$, have velocity $\vec{v}_{i,k} = (v_{i,k,x}, v_{i,k,y}, v_{i,k,z})$, and have acceleration $\vec{a}_{i,k} = (a_{i,k,x}, a_{i,k,y}, a_{i,k,z})$. The leap frog formulas, which relate position, velocity and acceleration are

$$\vec{v}_{i,\frac{1}{2}} = \vec{v}_{i,0} + \frac{1}{2}h\vec{a}_{i,0}, \quad \text{(Starter)} \tag{4.7}$$

$$\vec{v}_{i,k+\frac{1}{2}} = \vec{v}_{i,k-\frac{1}{2}} + h\vec{a}_{i,k}, \quad k = 1, 2, 3, \ldots. \tag{4.8}$$

$$\vec{r}_{i,k+1} = \vec{r}_{i,k} + h\vec{v}_{i,k+\frac{1}{2}}, \quad k = 0, 1, 2, 3, \ldots. \tag{4.9}$$

In formulas (4.7)–(4.9), the values $\vec{a}_{i,k}$ are determined from

$$\vec{a}_{i,k} = \frac{\vec{F}_{i,k}}{m_i}.$$

Note that in later discussions, confusion in notation will be avoided by using either h or Δt to represent a time step in a calculation.

A generic program for the leap frog formulas is as follows.

Algorithm 6 Program Leap Frog

Step 1. Set a time step h.
Step 2. Let distinct times t_k be given by $t_k = kh$, $k = 0, 1, 2, \ldots$
Step 3. Let $P(I)$ of mass $m(I)$ be N given particles, $I = 1, 2, \ldots, N$.
Step 4. For each I, let $P(I)$ be initially at $x(I, 0)$, $y(I, 0)$, $z(I, 0)$ with velocity components $vx(I, 0)$, $vy(I, 0)$, $vz(I, 0)$.
Step 5. For each I, let the force on $P(I)$ at any time t be $(Fx(I,t), Fy(I,t), Fz(I,t))$ and let the acceleration be $(Ax(I,t), Ay(I,t), Az(I,t))$, with
$$Ax(I,t) = Fx(I,t)/m(I)$$
$$Ay(I,t) = Fy(I,t)/m(I)$$
$$Az(I,t) = Fz(I,t)/m(I).$$
Step 6. For each I apply the starter formulas
$$vx(I, 0.5) = vx(I, 0) + \tfrac{1}{2}hAx(I, 0)$$
$$vy(I, 0.5) = vy(I, 0) + \tfrac{1}{2}hAy(I, 0)$$
$$vz(I, 0.5) = vz(I, 0) + \tfrac{1}{2}hAz(I, 0).$$
Step 7. For each I, determine positions and velocities sequentially by
$$x(I, t_k + h) = x(I, t_k) + (h)vx(I, t_k + 0.5h), \quad k = 0, 1, 2, \ldots$$
$$y(I, t_k + h) = y(I, t_k) + (h)vy(I, t_k + 0.5h), \quad k = 0, 1, 2, \ldots$$
$$z(I, t_k + h) = z(I, t_k) + (h)vz(I, t_k + 0.5h), \quad k = 0, 1, 2, \ldots$$
$$vx(I, t_p + 0.5h) = vx(I, t_p - 0.5h) + (h)Ax(I, t_p), \quad p = k+1.$$
$$vy(I, t_p + 0.5h) = vy(I, t_p - 0.5h) + (h)Ay(I, t_p), \quad p = k+1.$$
$$vz(I, t_p + 0.5h) = vz(I, t_p - 0.5h) + (h)Az(I, t_p), \quad p = k+1.$$
Step 8. Stop when k is sufficiently large.

4.6 Equations of Motion for Argon Vapor

From the discussion in Section 4.3, it follows that the equation of motion for a single argon vapor atom P_i acted on by a single argon vapor atom P_j, $i \neq j$, is

$$m\vec{a}_i = \left[\frac{209.0}{r_{ij}^{13}} - \frac{0.06551}{r_{ij}^{7}}\right] \frac{\vec{r}_{ji}}{r_{ij}} \quad \left(\frac{\text{g cm}}{\text{s}^2}\right). \tag{4.10}$$

Since the mass of an argon atom is $(6.63)10^{-23}$ g, and since $r = 10^8 R$, the latter equation is equivalent to

$$\vec{a}_i = \frac{10^{23}}{6.63}\left[\frac{209.0}{r_{ij}^{13}} - \frac{0.06551}{r_{ij}^{7}}\right] \frac{\vec{r}_{ji}}{r_{ij}} \quad \left(\frac{\text{cm}}{\text{s}^2}\right).$$

Replacing centimeters by angstroms and seconds by picoseconds (1 s $= 10^{12}$ ps), yields readily

$$\vec{a}_i = 10^6 \left[\frac{315.2}{r_{ij}^{13}} - \frac{0.09881}{r_{ij}^7} \right] \frac{\vec{r}_{ji}}{r_{ij}} \quad \left(\frac{\text{Å}}{\text{ps}^2} \right) \quad (4.11)$$

or, the second-order ordinary differential equation,

$$\frac{d^2 \vec{r}_i}{dT^2} = 98810 \left[\frac{3190}{r_{ij}^{13}} - \frac{1}{r_{ij}^7} \right] \frac{\vec{r}_{ji}}{r_{ij}} \quad \left(\frac{\text{Å}}{\text{ps}^2} \right) \quad (4.12)$$

which we will use.

On the molecular level, however, the effective force on P_i is local, that is it is determined only by close molecules. For argon atoms we choose the local interaction distance to be $D = 2.5\sigma = 2.5(3.418) = 8.545$ Å.

From (4.12), then, the dynamical equation for argon vapor atom P_i will be

$$\frac{d^2 \vec{r}_i}{dT^2} = 98810 \left[\frac{3190}{r_{ij}^{13}} - \frac{1}{r_{ij}^7} \right] \frac{\vec{r}_{ji}}{r_{ij}}; \quad r_{ij} < D. \quad (4.13)$$

The equations of motion for a system of N argon vapor atoms are then

$$\frac{d^2 \vec{r}_i}{dT^2} = 98810 \sum_{\substack{j \\ j \neq i}} \left[\frac{3190}{r_{ij}^{13}} - \frac{1}{r_{ij}^7} \right] \frac{\vec{r}_{ji}}{r_{ij}}; \quad i = 1, 2, 3, \ldots, N; \; r_{ij} < D. \quad (4.14)$$

Observe that on the molecular level gravity can be neglected since $980 \text{ cm/s}^2 = (980)10^{-16}$ Å/ps^2.

4.7
A Cavity Problem

In two dimensions consider now the square of side l shown in Figure 4.2. The interior is called the cavity or the basin of the square. The sides AB, BC, CD, DA are called the walls. Let argon vapor fill the basin. The top wall CD, alone, is allowed to move. It moves in the X direction at a constant speed V, called the wallspeed. Also it is allowed an extended length so that the fluid is always completely enclosed by four walls. (For the reasonableness of this assumption, see the treadmill apparatus of Koseff and Street (1984)). Then the cavity problem is to describe the gross motion of the fluid for various choices of V, which will be given in Å/ps.

Consider the cavity problem with $l = 230.22$ Å. Note that for a regular triangle with edge length 3.837 Å, which is the equilibrium distance, the altitude

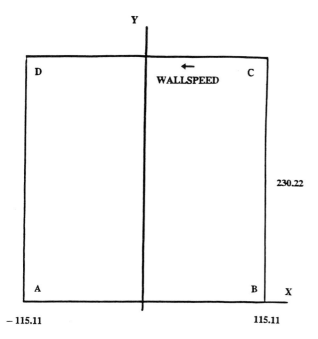

Fig. 4.2 Cavity.

is 3.323 Å. Using these values we now construct a regular triangular grid of 4235 points on the basin as follows:

$$x(1) = -115.11, \quad y(1) = 0$$
$$x(i) = x(i-1) + 3.837, \quad y(i) = 0, \quad i = 2, 61$$
$$x(62) = -113.1915, \quad y(62) = 3.323,$$
$$x(i) = x(i-1) + 3.837 \quad y(i) = 3.323, \quad i = 63, 121$$
$$x(i) = x(i-121), \quad y(i) = y(i-121) + 6.646, \quad i = 122, 4235.$$

At each point $(x(i), y(i))$ we set an argon atom P_i, $i = 1, 4235$. This array is shown in Figure 4.3 and provides the initial positions of the atoms.

To complete the initial data we need the initial velocities of the atoms. Let us choose the temperature to be 35 °C, so that $v = 3.58$ Å/ps. Each atom is then assigned a speed of 3.58 Å/ps in either the X or Y direction, determined at random, with its sign ($\pm$) also determined at random. Once the wallspeed V is prescribed, we are ready to solve the resulting cavity problem using the leap frog formulas for equations (4.14) with $N = 4235$. Note that the number of equations is 8470.

Fig. 4.3 Array of 4235 atoms.

4.8
Computational Considerations

For time step Δt (ps), and $t_k = k\Delta t$, two problems must be considered relative to the computations. The first problem is to prescribe a protocol when, computationally, an atom has crossed a wall into the exterior of the cavity. For each of the lower three walls, we will proceed as follows (no slip condition). The position will be reflected back symmetrically, relative to the wall, into the interior of the basin, the velocity component tangent to the wall will be set to zero and the velocity component perpendicular to the wall will be multiplied by -1. If the atom has crossed the moving wall, then its position will be reflected back symmetrically, its Y component of velocity will be multiplied by -1, and its X component of velocity will be increased by the wallspeed V.

The second problem derives from the fact that an instantaneous velocity field for molecular motion is Brownian. In order to better interpret gross fluid motion, we will introduce average velocities as follows. For J a positive integer, let particle P_i be at $(x(i,k), y(i,k))$ at t_k and at $(x(i, k-J), y(i, k-J))$ at t_{k-J}. Then the average velocity $\vec{v}_{i,k,J}$ of P_i at t_k is defined by

$$\vec{v}_{i,k,J} = \left(\frac{x(i,k) - x(i, k-J)}{J\Delta t}, \frac{y(i,k) - y(i, k-J)}{J\Delta t} \right). \quad (4.15)$$

4.9
Examples of Primary Vortex Generation

Consider first the parameter choices $V = -50$, $J = 25\,000$, $\Delta t = 0.000\,02$. Figures 4.4–4.7 show the development of a primary vortex at the respective times $T = 0.5, 1.0, 2.0, 3.5$. An initial compression wave is evident in Figure 4.4. The resulting compression in the lower right corner results in an upward repulsive effect which initiates the development of the counterclockwise vortex shown in Figure 4.7. The mean of the squares of the speeds at $T = 3.5$ is 156 Å/ps. Other values of J which were studied were 20 000, 15 000, and 10 000, each of which yielded results similar to those of Figures 4.4–4.7

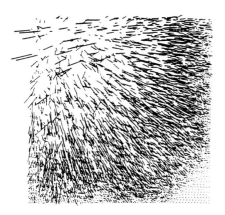

Fig. 4.4 $V = -50$, $J = 25000$, $T = 0.5$.

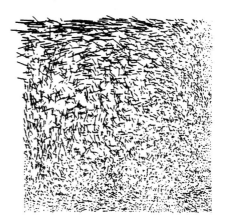

Fig. 4.5 $V = -50$, $J = 25000$, $T = 1.0$.

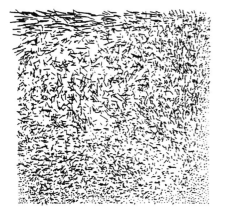

Fig. 4.6 $V = -50$, $J = 25000$, $T = 2.0$.

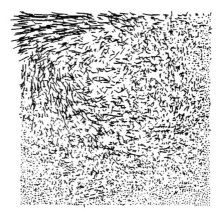

Fig. 4.7 $V = -50$, $J = 25000$, $T = 3.5$.

For the parameter choices $V = -100$, $J = 15\,000$, $\Delta t = 0.00\,002$, Figures 4.8–4.11 show the development of a primary vortex at the respective times $t = 0.5$, 1.5, 2.0, 2.5. An initial compression wave is seen in Figure 4.8 and again the compression in the lower right corner results in an upward repulsive effect which yield the counterclockwise vortex in Figure 4.11. The vortex in Figure 4.11 is larger than that in Figure 4.7 and has developed more quickly. The mean of the squares of the speeds at $T = 2.5$ is 222 Å/ps. Other values of J which were studied were 25 000, 20 000, and 10 000, each of which yielded results similar to those of Figures 4.8–4.11.

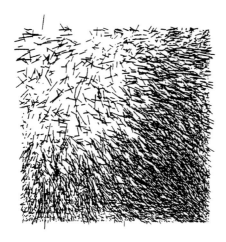

Fig. 4.8 $V = -100$, $J = 15000$, $T = 0.5$.

Fig. 4.9 $V = -100$, $J = 15000$, $T = 1.5$.

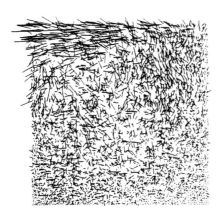

Fig. 4.10 $V = -100$, $J = 15000$, $T = 2.0$.

Fig. 4.11 $V = -100$, $J = 15000$, $T = 2.5$.

Other choices of V were -25 and -250, which yielded the expected results. For $V = -25$ the resulting primary vortex was smaller than that in Figure 4.7. For $V = -250$, the resulting primary vortex was larger than that in Figure 4.11.

4.10 Examples of Turbulent Flow

It is known that, in the large, turbulent flows have two well-defined criteria:

(1) A strong current develops across the usual primary vortex direction (Kolmogorov (1964)).

(2) Many small vortices appear and disappear quickly (Schlichting (1960)).

We will examine these criteria in our molecular calculations.

We take the following approach to generating turbulent flows. For a sufficiently large magnitude of the wallspeed V, let us show that turbulence results when, for given Δt, a stable calculation results but no J exists which yields a primary vortex.

Let us then set $V = -3000$, $\Delta t = 4(10)^{-7}$. The motion was simulated to $T = 1.11$. Typical results are shown at $T = 0.03, 0.09, 0.18, 0.27, 0.45, 0.72, 1.08,$ and 1.11 in Figures 4.12–4.19, respectively. Figure 4.12 shows the very rapid development of a compression wave. Figures 4.13–4.15 show what appears to be a large primary vortex. However, Figures 4.16–4.19 show the development of a strong vertical current in the left portion of the figure. Figure 4.19 shows that this current runs across the usual primary vortex direction not only in the left portion of the figure but, to a lesser degree, also in the right portion. Thus, Criterion (1) is valid.

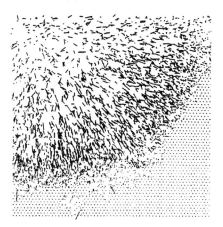
Fig. 4.12 $V = -3000$, $J = 75000$, $T = 0.03$.

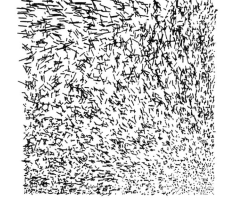
Fig. 4.13 $V = -3000$, $J = 75000$, $T = 0.09$.

We next define the concept of a *small vortex*. For $3 \leq M \leq 6$, we define a small vortex as a flow in which M molecules nearest to an $(M+1)$st molecule rotate either clockwise or counterclockwise about the $(M+1)$st molecule and, in addition, the $(M+1)$st molecule lies *interior* to a simple polygon determined by the given M molecules. With this definition, Figure 4.20 shows that

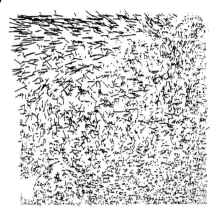

Fig. 4.14 $V = -3000$, $J = 75000$, $T = 0.18$.

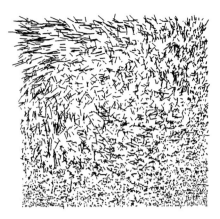

Fig. 4.15 $V = -3000$, $J = 75000$, $T = 0.27$.

Fig. 4.16 $V = -3000$, $J = 75000$, $T = 0.45$.

Fig. 4.17 $V = -3000$, $J = 75000$, $T = 0.72$.

the flow in Figure 4.18 has 198 small vortices at $T = 1.08$, while Figure 4.21 shows that at $T = 1.11$, only 0.03 picoseconds later, the resulting flow has 236 small vortices which are completely different than those from Figure 4.20.

Figures 4.22–4.24 show the same general flow as in Figure 4.19, but using $J = 60\,000, 45\,000$, and $30\,000$, respectively, so that the choice of J is immaterial to the development of turbulence.

Fig. 4.18 $V = -3000$, $J = 75000$, $T = 1.08$.

Fig. 4.19 $V = -3000$, $J = 75000$, $T = 1.11$.

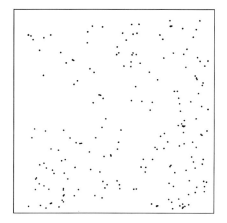

Fig. 4.20 198 small vortices in Figure 4.17.

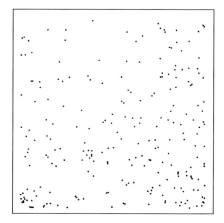

Fig. 4.21 236 small vortices in Figure 4.18.

4.11
Remark

Note now that it is known that nano mechanical results may differ from those in the large. This is indeed exactly the case in our computations, for Criterion 2 is also valid for primary vortex motions, like those shown in Figures 4.4–4.11.

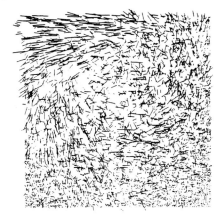
Fig. 4.22 $V = -3000$, $J = 60000$, $T = 1.11$.

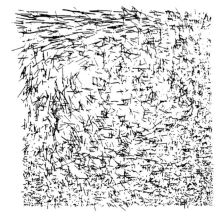
Fig. 4.23 $V = -3000$, $J = 45000$, $T = 1.11$.

Fig. 4.24 $V = -3000$, $J = 30000$, $T = 1.11$.

4.12
Molecular Formulas for Air

It is rather interesting that even though air is heterogeneous and consists of a variety of atoms and molecules, experimental Lennard–Jones potentials are readily available only for homogeneous air (Hirschfelder et al. (1967)). One such potential is

$$\phi(r_{ij}) = (5.36)10^{-14} \left[\frac{3.617^{12}}{r_{ij}^{12}} - \frac{3.617^{6}}{r_{ij}^{6}} \right] \text{ erg,} \tag{4.16}$$

in which r_{ij} is measured in angstroms. The force $\vec{F}_{ij}$ exerted on P_i by P_j is then

$$\vec{F}_{ij} = (5.36)10^{-6}\left[\frac{12(3.617^{12})}{r_{ij}^{13}} - \frac{6(3.617^6)}{r_{ij}^{7}}\right]\frac{\vec{r}_{ji}}{r_{ij}} \text{ dynes} \qquad (4.17)$$

and the equilibrium distance is $r_{ij} = 4.06$ Å. On the molecular level, the effective force on P_i is local, so that only molecules within a distance $D = 2.5\sigma = 2.5(3.617)$ Å are considered.

Before proceeding to dynamical considerations, it is necessary to characterize carefully the hypothetical air molecule to be used. We assume that the air to be used is non dilute and dry. Dry air (Masterton and Slowinski (1969)) consists primarily of 78% N_2, 21% O_2, and 1% Ar, whose respective masses are

$$m(N_2) = 28(1.660)10^{-24} \text{ g}$$
$$m(O_2) = 32(1.660)10^{-24} \text{ g}$$
$$m(Ar) = 40(1.660)10^{-24} \text{ g}.$$

We now characterize an "air" molecule A as consisting of proportionate amounts of N_2, O_2, and Ar and having mass

$$\begin{aligned} m(A) &= [0.78(28) + 0.21(32) + 0.01(40)](1.660)10^{-24} \\ &= (4.807)10^{-23} \text{ g}. \end{aligned} \qquad (4.18)$$

From (4.17) and (4.18) it follows that the acceleration, in Å/(ps)², of an air molecule P_i due to interaction with an air molecule P_j satisfies the equation

$$\vec{a}_i = (149795.)\left[\frac{4478}{r_{ij}^{13}} - \frac{1}{r_{ij}^{7}}\right]\frac{\vec{r}_{ji}}{r_{ij}}\left(\frac{\text{Å}}{\text{ps}^2}\right); \quad r_{ij} < D, \qquad (4.19)$$

in which we choose $D = 4.50$ Å. The equations of motion for a system of air molecules are then

$$\frac{d^2\vec{r}_i}{dT^2} = (149795.)\sum_{\substack{j \\ j\neq i}}\left[\frac{4478}{r_{ij}^{13}} - \frac{1}{r_{ij}^{7}}\right]\frac{\vec{r}_{ji}}{r_{ij}}\left(\frac{\text{Å}}{\text{ps}^2}\right);$$

$$i = 1, 2, 3, \ldots, N; \; r_{ij} < D. \quad (4.20)$$

4.13 A Cavity Problem

We consider again a cavity problem. In two dimensions consider the square of side 243.6 Å shown in Figure 4.25. The interior is the cavity or the basin.

The sides are the walls. Let the basin be filled with air. The top wall, alone, is allowed to move in the X direction with wallspeed V. The cavity problem is to describe the gross motion of the fluid for various choices of V, which will be given in Å/ps.

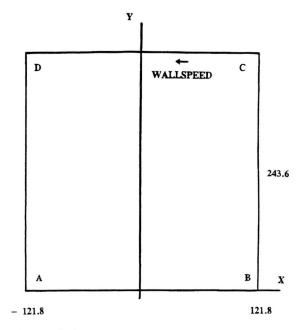

Fig. 4.25 Cavity.

4.14
Initial Data

Note that for a regular triangle with edge length 4.06 Å, which is the equilibrium distance, the altitude is 3.516 Å. Using these values we now construct a regular triangular grid of 4235 points on the basin as follows:

$$x(1) = -121.8, \qquad y(1) = 0$$
$$x(i) = x(i-1) + 4.06, \qquad y(i) = 0, \quad i = 2, 61$$
$$x(62) = -119.77, \qquad y(62) = 3.516$$
$$x(i) = x(i-1) + 4.06, \qquad y(i) = 3.516, \quad i = 63, 121$$
$$x(i) = x(i-121), \qquad y(i) = y(i-121) + 7.032, \quad i = 122, 4235.$$

At each point $(x(i), y(i))$ we set an air molecule P_i, $i = 1, 4235$. This array is that shown in Figure 4.26 and provides the initial positions of the molecules.

Fig. 4.26 Array of 4235 melocules.

To complete the initial data we need the initial velocities of the molecules. In two dimensions at 35 °C the speed of an air molecule is 4.207 Å/ps. Each molecule is then assigned a speed of 4.207 Å/ps in either the X or Y direction, determined at random, with its sign ($\pm$) also determined at random. Once the wallspeed V is prescribed, we are ready to solve the resulting cavity problem using the leap frog formulas for equations (4.20) with $N = 4235$.

The computational considerations are those given in Section 4.7.

4.15
Examples of Primary Vortex Generation

Consider first the parameter choices $V = -50$, $J = 25\,000$, $\Delta t = 0.000\,02$. Figures 4.27–4.30 show the development of a primary vortex at the respective times $T = 0.5, 1.0, 2.0, 4.0$. An initial compression wave is evident in Figure 4.27. The resulting compression in the lower right corner results in an upward repulsive effect which initiates the development of the counterclockwise vortex shown in Figure 4.30. The mean of the squares of the speeds at $T = 4.0$ is 120 Å/ps. Other values of J which were studied were 20 000, 15 000, and 10 000, each of which yielded results similar to those of Figures 4.27–4.30.

Other choices of V were -25, -100 and -250, which yielded the expected results.

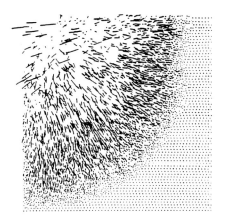

Fig. 4.27 $V = -50, J = 25000, T = 0.5$.

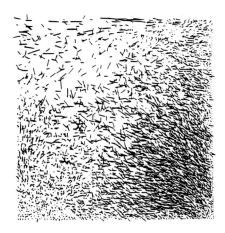

Fig. 4.28 $V = -50, J = 25000, T = 1.0$.

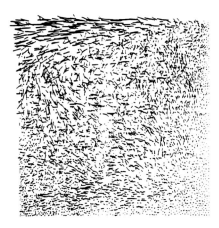

Fig. 4.29 $V = -50, J = 25000, T = 2.0$.

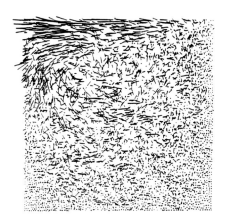

Fig. 4.30 $V = -50, J = 25000, T = 4.0$.

4.16
Turbulent Flow

We again take the following approach to generating turbulent flows. For a sufficiently large magnitude of the wallspeed V, let us show that turbulence results when, for given Δt, a stable calculation results, but no J exists which yields a primary vortex.

Let us then set $V = -3000$, $\Delta t = 4(10)^{-7}$, $J = 75000$. The motion was simulated to $T = 0.60$. Typical results are shown at $T = 0.09, 0.12, 0.21, 0.27, 0.36, 0.48, 0.57$, and 0.60 in Figures 4.31–4.38, respectively. Figure 4.31 shows the very rapid development of a compression wave. Figures 4.32–4.34 show what appears to be a large primary vortex. However, Figures 4.35–4.38 show the development of a strong vertical current in the left portion of each figure. Figure 4.38 shows that this current runs across the usual primary vortex direction not only in the left portion of the figure but, to a lesser degree, also in the right portion. Thus, Criterion (1) is valid.

Fig. 4.31 $V = -3000$, $J = 75000$, $T = 0.09$.

Fig. 4.32 $V = -3000$, $J = 75000$, $T = 0.12$.

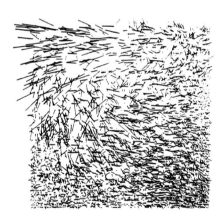

Fig. 4.33 $V = -3000$, $J = 75000$, $T = 0.21$.

Fig. 4.34 $V = -3000$, $J = 75000$, $T = 0.27$.

Fig. 4.35 $V = -3000$, $J = 75000$, $T = 0.36$.

Fig. 4.36 $V = -3000$, $J = 75000$, $T = 0.48$.

Fig. 4.37 $V = -3000$, $J = 75000$, $T = 0.57$.

Fig. 4.38 $V = -3000$, $J = 75000$, $T = 0.6$.

We again use the concept of a small vortex. Figure 4.39 shows that the flow in Figure 4.37 has 230 small vortices at $T = 0.57$, while Figure 4.40 shows that at $T = 0.60$, only 0.03 picoseconds later, the resulting flow has 240 small vortices which are completely different from those in Figure 4.39.

Figures 4.41–4.43 show the same general flow as in Figure 4.39, but using $J = 60\,000, 45\,000$, and $30\,000$, respectively, so that the choice of J is immaterial to the development of turbulence.

4.16 Turbulent Flow

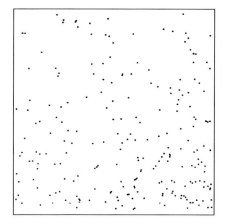

Fig. 4.39 230 small vortices in Figure 4.37.

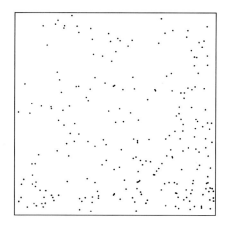

Fig. 4.40 240 small vortices in Figure 4.38.

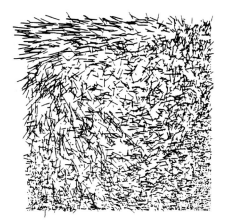

Fig. 4.41 $V = -3000$, $J = 60000$, $T = 0.60$.

Fig. 4.42 $V = -3000$, $J = 45000$, $T = 0.60$.

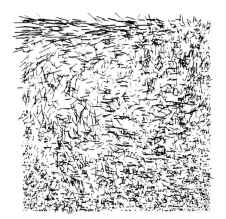

Fig. 4.43 $V = -3000$, $J = 30000$, $T = 0.60$.

4.17
Colliding Microdrops of Water Vapor

For N given water vapor molecules $P_1, P_2, \ldots, P_N$ which interact in accordance with classical molecular mechanics, the motion of each P_i can be approximated simplistically from given initial data by solving the differential system (Greenspan and Heath (1991)):

$$\frac{d^2 \vec{r}_i}{dT^2} = \sum_{\substack{j=1 \\ j \neq i}}^{N} \left(-\frac{16.5}{r_{ij}^3} + \frac{158.6}{r_{ij}^5} \right) \frac{\vec{r}_{ji}}{r_{ij}}, \quad i = 1, 2, \ldots, N, \qquad (4.21)$$

in which $\vec{r}_i$ is the position vector of P_i, $\vec{r}_{ji}$ is the vector from P_j to P_i, r_{ij} is the length of $\vec{r}_{ji}$, and T is a scaled time variable. The force in (4.21) is a least square approximation to a Lennard–Jones (6,12) potential (Hirschfelder et al. (1967)).

In order to simulate two colliding water vapor drops, it is convenient to generate first a single drop, which is done as follows. In the space cube $-31 \leq x \leq 31$, $-31 \leq y \leq 31$, $-31 \leq z \leq 31$, molecules are placed at the grid points generated by the choices $\Delta x = \Delta y = \Delta z = 3.1$. The space grid size 3.1 Å, is one that makes the term within the parenthesis in (4.21) equal to zero for $r = 3.1$. Next, the molecules outside the sphere whose equation is $x^2 + y^2 + z^2 = 26^2$ are deleted and the remaining molecules are assigned a random velocity in the range $|v_i| \leq 0.02$. At initial time, there are 2517 molecules and 7551 equations of form (4.21). This system is solved numerically by the leap frog formulas with $\Delta T = 0.0002$. The molecules are allowed to interact for 31 000 time steps. At T_{31000}, all molecules whose position coordinates satisfy $r > 26$ are deleted, reducing the number to 2051. The simulation is continued to T_{88000}, but with all velocities damped by a reset to zero at T_{40500}, T_{48000}, T_{53000}, T_{58000}, T_{63000}, T_{68000}, T_{73000}. At T_{78000}, the velocities are damped by the factor 0.5. No damping is imposed thereafter. The damping process cools the molecular configuration so that at T_{88000}, the temperature of the resulting drop is 45 °C. This drop is shown in Figure 4.44.

In order to study collisions, the single drop just generated is duplicated by mirror imaging. The resulting two drops are set symmetrically 3 Å apart about the YZ plane, as shown in Figure 4.45. To elucidate the motions of individual molecules during collision, the drops are displayed in different shadings. To avoid complete symmetry, the velocities of any molecule and its mirror image are taken to be the same. In addition, the counter is reset to zero.

To simulate collision, we assume that each molecule of the "light" drop, on the left in Figure 4.45, has its velocity increased by $\vec{V}$, while each molecule of "dark" drop has its velocity decreased by $\vec{V}$.

Consider first setting $\vec{V} = (0,0,0)$, so that the two drops interact with no changes in velocity. The numerical solution of the 12 306 second-order equa-

4.17 Colliding Microdrops of Water Vapor

tions is shown at T_{9500}, T_{18500}, T_{27500}, T_{36500} in Figure 4.46. The resulting mode of interaction is called an oscillating oblateness mode.

For the choice $\vec{V} = (2.2, 0.2, 0.0)$, Figure 4.47 shows, at the times T_{9500}, T_{18500}, T_{27500}, the development of a raindrop mode.

For the choice $\vec{V} = (5.0, 10.0, 0.0)$, Figure 4.48 shows a resulting non cohesive collision which exhibits a clean slicing effect, the molecular transfer during collision, and the loss of molecules after separation. The times shown are T_{4000}, T_{9000}, T_{13500}, T_{18000}.

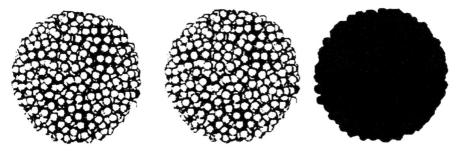

Fig. 4.44 A microdrop of water at 45°C.

Fig. 4.45 Two microdrops of water at 3 Å apart.

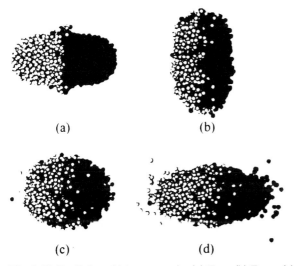

(a) (b)

(c) (d)

Fig. 4.46 Oscillating oblateness mode. (a) T_{9500}, (b) T_{18500}, (c) T_{27500}, (d) T_{36500}.

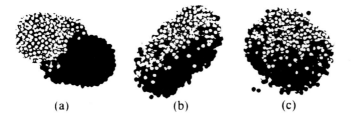

Fig. 4.47 Raindrop mode. (a) T_{9500}, (b) T_{18500}, (c) T_{27500}.

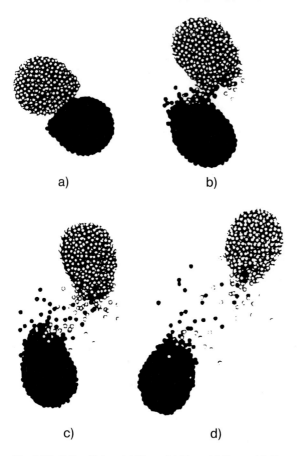

Fig. 4.48 Soft collision. (a) T_{4000}, (b) T_{9000}, (c) T_{13500}, (d) T_{18000}.

4.18
Remarks

Note that by using the formulas of this chapter it is possible to extend fluid flow and fragmentation applications into the large by means of the engineering lumped mass technique (Wang and Ostoja-Starzewski (2005); Greenspan

(2005)). Using conservation of mass and energy, one can develop a particle mechanics methodology at the large scale which is an extension of the molecular mechanics at the small scale.

Note also that the calculations described in Section 4.17 were executed originally on a Cray supercomputer in 1991. Today's scientific personal computers, like the Alpha 533, are as fast as that 1991 Cray and bring molecular computations into the realm of the reasonable.

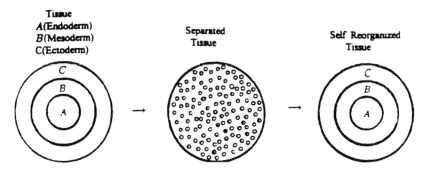

Fig. 4.49 The Holtfreter experiment.

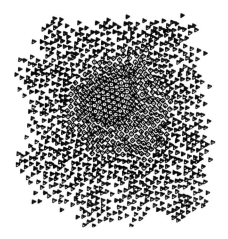

Fig. 4.50 Completed self reorganization.

Finally, let us show that the leap frog formulas, applied thus far only on the molecular level, are also of value in the large. An interesting area of biological study of wide interest is cellular self reorganization. This has been motivated by some intriguing experiments (Steinberg (1975)). For example, Holtfreter showed that embryonic tissue, consisting of distinct endoderm, mesoderm and ectoderm layers, when separated out, could recombine into tissue with normal endoderm, mesoderm and ectoderm layers (see Figure 4.49). As another example, in an experiment by Wilson, cells and cell clusters obtained by squeezing a sponge through a fine silk cloth could reunite and aggregates could reconstruct themselves into functional sponges. Figure 4.50 shows a leap frog solution and the final self reorganization of 1072 separated cells into endoderm, mesoderm and ectoderm layers (Greenspan (1997)).

4.19
Exercises

4.1 Consider two particles P_1 and P_2 in motion in the XY plane. Let P_1 have mass $m_1 = 2$ and initial data $x_{1,0} = 0$, $y_{1,0} = 10$, $v_{1,0,x} = 0$, $v_{1,0,y} = -15$. Let P_2 have mass $m_2 = 1$ and initial data $x_{2,0} = 10$, $y_{2,0} = 0$, $v_{2,0,x} = -10$, $v_{2,0,y} = -4$. Using the leap frog formulas with $h = 10^{-3}$, determine the motion of P_1 and P_2 through t_{1000} for the force formulas

$$\vec{F}_{1,k} = \left[-\frac{1}{(r_{12,k})^2} + \frac{4}{(r_{12,k})^4} \right] \frac{\vec{r}_{21,k}}{r_{12,k}}, \quad \vec{F}_{2,k} = -\vec{F}_{1,k}.$$

4.2 Repeat Exercise 1 but use the force formulas

$$\vec{F}_{1,k} = \left[-\frac{1}{(r_{12,k})^3} + \frac{1}{(r_{12,k})^5} \right] \frac{\vec{r}_{21,k}}{r_{12,k}}, \quad \vec{F}_{2,k} = -\vec{F}_{1,k}.$$

Write the program so that no square roots are required.

4.3 Consider two particles P_1 and P_2 in motion in three dimensions. Let P_1 have mass $m_1 = 2$ and P_2 have mass $m_2 = 1$. Let the initial data be

$x_{1,0} = -1$, $y_{1,0} = 0$, $z_{1,0} = 0$, $v_{1,0,x} = 0$, $v_{1,0,y} = 0$, $v_{1,0,z} = -1$
$x_{2,0} = 1$, $y_{2,0} = 0$, $z_{2,0} = 0$, $v_{2,0,x} = 0$, $v_{2,0,y} = 5$, $v_{2,0,z} = 1$.

Using the leap frog formulas with $h = 10^{-4}$, determine the motion of P_1 and P_2 through t_{10000} for the force formulas

$$\vec{F}_{1,k} = -\frac{m_1 m_2}{(r_{12,k})^2} \frac{\vec{r}_{21,k}}{r_{12,k}}, \quad \vec{F}_{2,k} = -\vec{F}_{1,k}.$$

Graph the resulting motion.

4.4 Consider three particles P_1, P_2, P_3, of respective masses $m_1 = 1000$, $m_2 = 20$, $m_3 = 1$. Let them be in motion in the XY plane. The initial data are

$x_{1,0} = 0$, $y_{1,0} = 0$, $v_{1,0,x} = 0$, $v_{1,0,y} = 0$
$x_{2,0} = 10$, $y_{2,0} = 0$, $v_{2,0,x} = 0$, $v_{2,0,y} = 10$
$x_{3,0} = -10$, $y_{3,0} = 0$, $v_{3,0,x} = 0$, $v_{3,0,y} = -5$.

Using the leap frog formulas with $h = 10^{-4}$, determine the motion of P_1, P_2 and P_3 through t_{10000} for the force formulas

$$\vec{F}_{1,k} = -\frac{m_1 m_2}{(r_{12,k})^2} \frac{\vec{r}_{21,k}}{r_{12,k}} - \frac{m_1 m_3}{(r_{13,k})^2} \frac{\vec{r}_{31,k}}{r_{13,k}}$$

$$\vec{F}_{2,k} = -\frac{m_1 m_2}{(r_{12,k})^2} \frac{\vec{r}_{12,k}}{r_{12,k}} - \frac{m_2 m_3}{(r_{23,k})^2} \frac{\vec{r}_{32,k}}{r_{23,k}}$$

$$\vec{F}_{3,k} = -\frac{m_1 m_3}{(r_{13,k})^2} \frac{\vec{r}_{13,k}}{r_{13,k}} - \frac{m_2 m_3}{(r_{23,k})^2} \frac{\vec{r}_{23,k}}{r_{23,k}}.$$

4.5 Reproduce Figure 4.7.

4.6 Reproduce Figure 4.11.

4.7 Reproduce Figure 4.19.

4.8 Reproduce Figure 4.48.

5
Completely Conservative, Covariant Numerical Methodology

5.1
Introduction

When the force between two particles of an N-body problem depends only on the distance between the particles, the system conserves energy, linear momentum and angular momentum. Also, the Newtonian dynamical equations are covariant, that is, they retain their structure under fundamental coordinate transformations. In this chapter, we will demonstrate how such systems can be solved by a numerical procedure which conserves exactly the same invariants as does the differential system. In addition, we will show that the approximating difference equations will also be covariant. In practice, the methodology is implicit and therefore less economical than other methods. However, when such quantities as energy must be conserved, the method is of value.

5.2
Mathematical Considerations

For clarity, we proceed in three dimensions with a fundamental N-body problem, that is, with $N = 3$. This problem contains all the complexities for larger values of N, whereas the case $N = 2$ does not. Extension to arbitrary $N > 3$ follows readily using entirely analogous proofs as those given for $N = 3$. Of course, the discussion applies equally well to two dimensions. Throughout, cgs units are used.

For $i = 1, 2, 3$, let P_i of mass m_i be at $\vec{r}_i = (x_i, y_i, z_i)$ at time t. Let the positive distance between P_i and P_j, $i \neq j$, be r_{ij}, with $r_{ij} = r_{ji}$. Let $\phi(r_{ij}) = \phi_{ij}$, given in ergs, be a potential for the pair P_i, P_j. Then the Newtonian dynamical

---

Numerical Solution of Ordinary Differential Equations for Classical, Relativistic and Nano Systems. Donald Greenspan
Copyright © 2006 WILEY-VCH Verlag GmbH & Co. KGaA, Weinheim
ISBN: 3-527-40610-7

equations for the three-body interaction are

$$m_i \frac{d^2 \vec{r}_i}{dt^2} = -\frac{\partial \phi_{ij}}{\partial r_{ij}} \frac{\vec{r}_i - \vec{r}_j}{r_{ij}} - \frac{\partial \phi_{ik}}{\partial r_{ik}} \frac{\vec{r}_i - \vec{r}_k}{r_{ik}}, \quad i = 1, 2, 3, \qquad (5.1)$$

where $j = 2$ and $k = 3$ when $i = 1$; $j = 1$ and $k = 3$ when $i = 2$; $j = 1$ and $k = 2$ when $i = 3$.

System (5.1) conserves energy, linear momentum, and angular momentum. In addition, it is covariant, that is, it has the same functional form under translation, rotation, and uniform relative motion of coordinate frames. Our problem is to devise a numerical scheme for solving system (5.1) from given initial data so that the numerical scheme is also covariant and preserves the same system invariants.

5.3
Numerical Methodology

For $h > 0$, let $t_n = nh$, $n = 0, 1, 2, \ldots$. At time t_n, let P_i be at $\vec{r}_{i,n} = (x_{i,n}, y_{i,n}, z_{i,n})$ with velocity $\vec{v}_{i,n} = (v_{i,n,x}, v_{i,n,y}, v_{i,n,z})$, and denote the distances $|P_1 P_2|, |P_1 P_3|, |P_2 P_3|$ by $r_{12,n}, r_{13,n}, r_{23,n}$ respectively. We now approximate the second-order differential system (5.1) by the first-order difference system

$$\frac{\vec{r}_{i,n+1} - \vec{r}_{i,n}}{h} = \frac{\vec{v}_{i,n+1} + \vec{v}_{i,n}}{2} \qquad (5.2)$$

$$m_i \frac{\vec{v}_{i,n+1} - \vec{v}_{i,n}}{h} = -\frac{\phi(r_{ij,n+1}) - \phi(r_{ij,n})}{r_{ij,n+1} - r_{ij,n}} \frac{\vec{r}_{i,n+1} + \vec{r}_{i,n} - \vec{r}_{j,n+1} - \vec{r}_{j,n}}{r_{ij,n+1} + r_{ij,n}} \qquad (5.3)$$
$$- \frac{\phi(r_{ik,n+1}) - \phi(r_{ik,n})}{r_{ik,n+1} - r_{ik,n}} \frac{\vec{r}_{i,n+1} + \vec{r}_{i,n} - \vec{r}_{k,n+1} - \vec{r}_{k,n}}{r_{ik,n+1} + r_{ik,n}}$$

where $j = 2$ and $k = 3$ when $i = 1$; $j = 1$ and $k = 3$ when $i = 2$; $j = 1$ and $k = 2$ when $i = 3$.

In all the problems to be considered, $\phi(r_{ij,n+1}) - \phi(r_{ij,n})$ will contain the factor $r_{ij,n+1} - r_{ij,n}$, so that the singularity $r_{ij,n+1} = r_{ij,n}$ is removable.

System (5.2)–(5.3) constitutes 18 implicit recursion equations for the unknowns $x_{i,n+1}, y_{i,n+1}, z_{i,n+1}, v_{i,n+1,x}, v_{i,n+1,y}, v_{i,n+1,z}$ in the 18 knowns $x_{i,n}, y_{i,n}, z_{i,n}, v_{i,n,x}, v_{i,n,y}, v_{i,n,z}$, $i = 1, 2, 3$. These equations can be solved readily by the following form of Newton's method (Greenspan (1980)) to yield the numerical solution. For the nonlinear algebraic or transcendental system

$$f_1(x_1, x_2, \ldots, x_k) = 0$$
$$f_2(x_1, x_2, \ldots, x_k) = 0$$
$$\vdots$$
$$f_k(x_1, x_2, \ldots, x_k) = 0,$$

iterate to convergence from an initial guess with

$$x_1^{(n+1)} = x_1^{(n)} - \frac{f_1(x_1^{(n)}, x_2^{(n)}, \ldots, x_k^{(n)})}{\frac{\partial f_1}{\partial x_1}(x_1^{(n)}, x_2^{(n)}, \ldots, x_k^{(n)})}$$

$$x_2^{(n+1)} = x_2^{(n)} - \frac{f_2(x_1^{(n)}, x_2^{(n)}, \ldots, x_k^{(n)})}{\frac{\partial f_2}{\partial x_2}(x_1^{(n)}, x_2^{(n)}, \ldots, x_k^{(n)})}$$

$$\vdots$$

$$x_k^{(n+1)} = x_k^{(n)} - \frac{f_k(x_1^{(n)}, x_2^{(n)}, \ldots, x_k^{(n)})}{\frac{\partial f_k}{\partial x_k}(x_1^{(n)}, x_2^{(n)}, \ldots, x_k^{(n)})}.$$

Often the initial guess $x_1^{(0)} = x_2^{(0)} = \ldots = x_k^{(0)} = 0.0$ is adequate. However, if convergence does not result, then a different initial guess must be chosen.

Since no proof has been provided that a solution of the equations exists, one must also provide *a posteriori* proof by substitution of the result of the convergent iteration into the original set of equations to show that it is a solution.

5.4
Conservation Laws

Let us show now that the numerical solution generated by (5.2)–(5.3) conserves the same energy, linear momentum, and angular momentum as does (5.1).

Consider first energy conservation. For this purpose, define

$$W_N = \sum_{n=0}^{N-1} \left\{ \sum_{i=1}^{3} m_i (\vec{r}_{i,n+1} - \vec{r}_{i,n}) \cdot (\vec{v}_{i,n+1} - \vec{v}_{i,n})/h \right\}. \quad (5.4)$$

Note immediately relative to (5.4) that, since we are considering specifically the three-body problem, the symbol N in summation (5.4) is now simply a numerical time index. Then insertion of (5.2) into (5.4) and simplification yields

$$W_N = \frac{1}{2} m_1 (v_{1,N})^2 + \frac{1}{2} m_2 (v_{2,N})^2 + \frac{1}{2} m_3 (v_{3,N})^2$$
$$- \frac{1}{2} m_1 (v_{1,0})^2 - \frac{1}{2} m_2 (v_{2,0})^2 - \frac{1}{2} m_3 (v_{3,0})^2,$$

so that
$$W_N = K_N - K_0. \tag{5.5}$$

Insertion of (5.3) into (5.4) implies, with some tedious but elementary algebraic manipulation (see Exercise 2),

$$W_N = \sum_{n=0}^{N-1} (-\phi_{12,n+1} - \phi_{13,n+1} - \phi_{23,n+1} + \phi_{12,n} + \phi_{13,n} + \phi_{23,n})$$

so that
$$W_N = -\phi_N + \phi_0. \tag{5.6}$$

Elimination of W_N between (5.5) and (5.6) then yields conservation of energy, that is,
$$K_N + \phi_N = K_0 + \phi_0, \quad N = 1, 2, 3, \ldots.$$

Moreover, since K_0 and ϕ_0 depend only on initial data, it follows that K_0 and ϕ_0 are the same in both the continuous and the discrete cases, so that the energy conserved by the numerical method is exactly that of the continuous system. Note, *in addition*, that the proof is independent of h. Thus, we have proved the following theorem.

Theorem 5.1 *Independently of h, the numerical method of Section 5.3 is energy conserving, that is,*

$$K_N + \phi_N = K_0 + \phi_0, \quad N = 1, 2, 3, \ldots.$$

Next, the linear momentum $\vec{M}_i(t_n) = \vec{M}_{i,n}$ of P_i at t_n is defined to be the vector
$$\vec{M}_{i,n} = m_i(v_{i,n,x}, v_{i,n,y}, v_{i,n,z}). \tag{5.7}$$

The linear momentum $\vec{M}_n$ of the three-body system at time t_n is defined to be the vector
$$\vec{M}_n = \sum_{i=1}^{3} \vec{M}_{i,n}. \tag{5.8}$$

Now, from (5.3),
$$m_1(\vec{v}_{1,n+1} - \vec{v}_{1,n}) + m_2(\vec{v}_{2,n+1} - \vec{v}_{2,n}) + m_3(\vec{v}_{3,n+1} - \vec{v}_{3,n}) \equiv \vec{0}. \tag{5.9}$$

Thus, in particular, for $n = 0, 1, 2, \ldots$,
$$m_1(v_{1,n+1,x} - v_{1,n,x}) + m_2(v_{2,n+1,x} - v_{2,n,x}) + m_3(v_{3,n+1,x} - v_{3,nx}) = 0. \tag{5.10}$$

Summing both sides of (5.10) from $n = 0$ to $n = N - 1$ implies
$$m_1 v_{1,N,x} + m_2 v_{2,N,x} + m_3 v_{3,N,x} = C_1, \quad N \geq 1 \tag{5.11}$$

5.4 Conservation Laws

in which
$$m_1 v_{1,0,x} + m_2 v_{2,0,x} + m_3 v_{3,0,x} = C_1. \tag{5.12}$$

Similarly,
$$m_1 v_{1,N,y} + m_2 v_{2,N,y} + m_3 v_{3,N,y} = C_2 \tag{5.13}$$
$$m_1 v_{1,N,z} + m_2 v_{2,N,z} + m_3 v_{3,N,z} = C_3 \tag{5.14}$$

in which
$$m_1 v_{1,0,y} + m_2 v_{2,0,y} + m_3 v_{3,0,y} = C_2 \tag{5.15}$$
$$m_1 v_{1,0,z} + m_2 v_{2,0,z} + m_3 v_{3,0,z} = C_3. \tag{5.16}$$

Thus,
$$\vec{M}_n = \sum_{i=1}^{3} \vec{M}_{i,n} = (C_1, C_2, C_3) = \vec{M}_0, \quad n = 1, 2, 3, \ldots,$$

which is the classical law of conservation of linear momentum. Note that $\vec{M}_0$ depends only on the initial data. Thus we have the following theorem.

Theorem 5.2 *Independently of h, the numerical method of Section 5.3 conserves linear momentum, that is,*
$$\vec{M}_n = \vec{M}_0, \quad n = 1, 2, 3, \ldots.$$

We turn finally to angular momentum. The angular momentum $\vec{L}_{i,n}$ of P_i at t_n is defined to be the cross product vector
$$\vec{L}_{i,n} = m_i (\vec{r}_{i,n} \times \vec{v}_{i,n}). \tag{5.17}$$

The angular momentum of a three-body system at t_n is defined to be the vector
$$\vec{L}_n = \sum_{i=1}^{3} \vec{L}_{i,n}.$$

It then follows readily (Greenspan (1980)) that
$$\vec{L}_{n+1} - \vec{L}_n = m_i \sum_{i=1}^{3} \left[(\vec{r}_{i,n+1} - \vec{r}_{i,n}) \times \frac{1}{2} (\vec{v}_{i,n+1} + \vec{v}_{i,n}) \right. \\ \left. + \frac{1}{2} (\vec{r}_{i,n+1} + \vec{r}_{i,n}) \times (\vec{v}_{i,n+1} - \vec{v}_{i,n}) \right]. \tag{5.18}$$

However, substitution of (5.3) into (5.18) yields, after some tedious calculations using the basic laws of vector products (Exercise 4),
$$\vec{L}_{n+1} - \vec{L}_n = \vec{0}, \quad n = 0, 1, 2, 3, \ldots,$$

so that
$$\vec{L}_n = \vec{L}_0, \quad n = 1, 2, 3, \ldots,$$

which implies, independently of h, the conservation of angular momentum. Note again that $\vec{L}_0$ depends only on the initial data. Thus the following theorem has been proved.

Theorem 5.3 *Independently of h, the numerical method of Section 5.3 conserves angular momentum, that is*

$$\vec{L}_n = \vec{L}_0, \quad n = 1, 2, 3, \ldots.$$

5.5
Covariance

Let us begin the discussion of covariance as simply as possible. When a dynamical equation is structurally invariant under a transformation, the equation is said to be covariant or symmetric. The transformations we will consider are translation, rotation, and uniform relative motion. We will concentrate on two-dimensional systems, because the related techniques and results extend directly to three dimensions. A general Newtonian force will be considered, so that the particular type of force discussed in Sections 5.2–5.4 will be included as a special case. Finally, we will concentrate on the motion of a single particle P of mass m, with extension to the N-body problem following in a natural way. And though the assumptions just made may seem to be excessive, it will be seen shortly that they make the required mathematical methodology readily transparent.

Suppose now that a particle P of mass m is in motion in the XY plane and that for $\Delta t = h > 0$ its motion from given initial data is determined by a force $\vec{F}(t_n) = \vec{F}_n = (F_{n,x}, F_{n,y})$ and by the dynamical difference equations

$$F_{n,x} = m(v_{n+1,x} - v_{n,x})/h \tag{5.19}$$

$$F_{n,y} = m(v_{n+1,y} - v_{n,y})/h. \tag{5.20}$$

Our problem is as follows. Let $x = f_1(x^*, y^*)$, $y = f_2(x^*, y^*)$ be a change of coordinates. Under this transformation, let $F_{n,x} = F^*_{n,x^*}$, $F_{n,y} = F^*_{n,y^*}$. Then we will want to prove that in the X^*Y^* system the dynamical equations of motion are

$$F^*_{n,x^*} = m(v_{n+1,x^*} - v_{n,x^*})/h \tag{5.21}$$

$$F^*_{n,y^*} = m(v_{n+1,y^*} - v_{n,y^*})/h, \tag{5.22}$$

which will establish covariance.

In consistency with (5.2), we assume that

$$\frac{x_{n+1} - x_n}{h} = \frac{v_{n+1,x} + v_{n,x}}{2}, \quad \frac{x^*_{n+1} - x^*_n}{h} = \frac{v_{n+1,x^*} + v_{n,x^*}}{2} \tag{5.23}$$

$$\frac{y_{n+1} - y_n}{h} = \frac{v_{n+1,y} + v_{n,y}}{2}, \quad \frac{y^*_{n+1} - y^*_n}{h} = \frac{v_{n+1,y^*} + v_{n,y^*}}{2}. \tag{5.24}$$

Relative to (5.23) and (5.24), the following lemma will be of value.

5.5 Covariance

Lemma 5.1 *Equations* (5.23) *and* (5.24) *imply*

$$v_{1,x} = \frac{2}{h}(x_1 - x_0) - v_{0,x}; \quad v_{1,x^*} = \frac{2}{h}(x_1^* - x_0^*) - v_{0,x^*} \quad (5.25)$$

$$v_{1,y} = \frac{2}{h}(y_1 - y_0) - v_{0,y}; \quad v_{1,y^*} = \frac{2}{h}(y_1^* - y_0^*) - v_{0,y^*} \quad (5.26)$$

$$v_{n,x} = \frac{2}{h}\left[x_n + (-1)^n x_0 + 2\sum_{j=1}^{n-1}(-1)^j x_{n-j}\right] + (-1)^n v_{0,x}, \quad n \geq 2 \quad (5.27a)$$

$$v_{n,x^*} = \frac{2}{h}\left[x_n^* + (-1)^n x_0^* + 2\sum_{j=1}^{n-1}(-1)^j x_{n-j}^*\right] + (-1)^n v_{0,x^*}, \quad n \geq 2 \quad (5.27b)$$

$$v_{n,y} = \frac{2}{h}\left[y_n + (-1)^n y_0 + 2\sum_{j=1}^{n-1}(-1)^j y_{n-j}\right] + (-1)^n v_{0,y}, \quad n \geq 2 \quad (5.28a)$$

$$v_{n,y^*} = \frac{2}{h}\left[y_n^* + (-1)^n y_0^* + 2\sum_{j=1}^{n-1}(-1)^j y_{n-j}^*\right] + (-1)^n v_{0,y^*}, \quad n \geq 2. \quad (5.28b)$$

Proof. Equations (5.25) follow directly from (5.23) with $n = 0$. Equations (5.26) follow directly from (5.24) with $n = 0$. Equations (5.27a), (5.27b), (5.28a), (5.28b) follow readily by mathematical induction (Exercise 5). □

Theorem 5.4 *Equations* (5.19), (5.20) *are covariant relative to the translation*

$$x^* = x - a, \quad y^* = y - b; \quad a, b \text{ constants}. \quad (5.29)$$

Proof. Define $v_{0,x} = v_{0,x^*}$, $v_{0,y} = v_{0,y^*}$. Then, from (5.25),

$$v_{1,x} = \frac{2}{h}[(x_1^* + a) - (x_0^* + a)] - v_{0,x^*} = v_{1,x^*}.$$

Similarly,

$$v_{1,y} = v_{1,y^*}.$$

For $n > 1$, (5.27a) and (5.27b) yield

$$v_{n,x} = \frac{2}{h}\left[(x_n^* + a) + (-1)^n(x_0^* + a) + 2\sum_{j=1}^{n-1}(-1)^j(x_{n-j}^* + a)\right] + (-1)^n v_{0,x^*}. \quad (5.30)$$

However, by the lemma, for n both odd and even, (5.30) implies

$$v_{n,x} = v_{n,x^*}. \quad (5.31)$$

Similarly,

$$v_{n,y} = v_{n,y^*}. \quad (5.32)$$

Thus, for all $n = 0, 1, 2, 3, \ldots$

$$v_{n,x} = v_{n,x^*}$$
$$v_{n,y} = v_{n,y^*}.$$

Thus,

$$F^*_{n,x^*} = F_{n,x} = m \frac{v_{n+1,x} - v_{n,x}}{h} = m \frac{v_{n+1,x^*} - v_{n,x^*}}{h}.$$

Similarly,

$$F^*_{n,y^*} = m \frac{v_{n+1,y^*} - v_{n,y^*}}{h},$$

and the theorem is proved. □

Theorem 5.5 *Under the rotation*

$$\left.\begin{array}{l} x^* = x \cos \theta + y \sin \theta \\ y^* = y \cos \theta - x \sin \theta \end{array}\right\} \quad (5.33)$$

where θ is the smallest positive angle measured counterclockwise from the X to the X^* axis, equations (5.19) and (5.20) are covariant.

Proof. The proof follows along the same lines as that of Theorem 5.1 (Exercise 7) after one defines

$$\left.\begin{array}{l} v_{0,x^*} = v_{0,x} \cos \theta + v_{0,y} \sin \theta \\ v_{0,y^*} = v_{0,y} \cos \theta - v_{0,x} \sin \theta \end{array}\right\} \quad (5.34)$$

and observes that

$$\left.\begin{array}{l} F^*_{n,x^*} = F_{n,x} \cos \theta + F_{n,y} \sin \theta \\ F^*_{n,y^*} = F_{n,y} \cos \theta - F_{n,x} \sin \theta. \end{array}\right\} \quad (5.35)$$

□

Theorem 5.6 *Under relative uniform motion of coordinate systems, equations (5.19), (5.20) are covariant.*

Proof. Consider first motion in one dimension. Assume then that the X and X^* axes are in relative motion defined by

$$x^*_n = x_n - ct_n, \quad n = 0, 1, 2, 3, \ldots, \quad (5.36)$$

in which c is a positive constant. If $v_{0,x}$ is the initial velocity of P along the X axis, let v_{0,x^*} along the X^* axis be defined by

$$v_{0,x^*} = v_{0,x} - c. \quad (5.37)$$

Hence, for $n = 1$,
$$v_{1,x} = \frac{2}{h}[(x_1^* + ct_1) - (x_0^* + ct_0)] - v_{0,x} = v_{1,x^*} + c.$$

For $n \geq 2$,
$$v_{n,x} = \frac{2}{h}\{x_n^* + (-1)^n x_0^* + 2\sum_{j=1}^{n-1}(-1)^j x_{n-j}^*\} + (-1)^n v_{0,x}$$
$$+ \frac{2c}{h}\{t_n + (-1)^n t_0 + 2\sum_{j=1}^{n-1}(-1)^j t_{n-j}\}.$$

But,
$$t_n + (-1)^n t_0 + 2\sum_{j=1}^{n-1}(-1)^j t_{n-j} = \begin{cases} 0, & n \text{ even} \\ h, & n \text{ odd.} \end{cases}$$

Thus, with the aid of the lemma, it follows that for both n odd and even,
$$v_{n,x} = v_{n,x^*} + c.$$

Thus for all $n = 0, 1, 2, 3, \ldots$,
$$F_{n,x^*}^* = F_{n,x} = m\frac{v_{n+1,x^*} + c - v_{n,x^*} - c}{h} = m\frac{v_{n+1,x^*} - v_{n,x^*}}{h}.$$

Under the assumption that
$$y^* = y - dt_n$$
in which d is a constant, one finds similarly that
$$F_{n,y^*}^* = m\frac{v_{n+1,y^*} - v_{n,y^*}}{h},$$
and the covariance is established. □

5.6
Application – A Spinning Top on a Smooth Horizontal Plane

Rigid body motion is of fundamental interest in mathematics, science, and engineering. In this section we introduce a new, simplistic approach to this area of study in the spirit of modern molecular mechanics. We will consider a discrete, or, lumped mass, tetrahedral body and simulate its motion when it spins like a top. The approach will not require the use of special coordinates, Cayley–Klein parameters, tensors, dyadics, or related concepts. All that will be required is Newtonian mechanics in three-dimensional XYZ space. The numerical methodology will conserve exactly the same energy, linear momentum, and angular momentum as does the associated differential system.

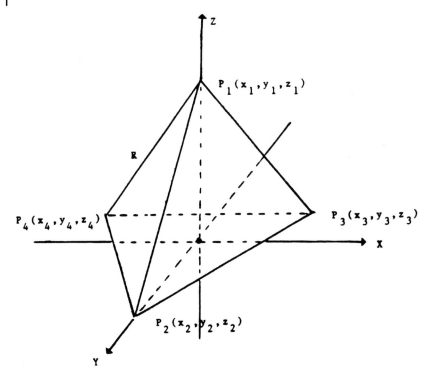

Fig. 5.1 A regular tetrahedron.

Consider, as shown in Figure 5.1, a regular tetrahedron with vertices $P_i(x_i, y_i, z_i)$, $i = 1, 2, 3, 4$, and edge length R. For convenience set

$$(x_1, y_1, z_1) = (0, 0, R\sqrt{6}/3), \qquad (x_2, y_2, z_2) = (0, R\sqrt{3}/3, 0),$$

$$(x_3, y_3, z_3) = \left(\frac{1}{2}R, -R\sqrt{3}/6, 0\right), \quad (x_4, y_4, z_4) = \left(-\frac{1}{2}R, -R\sqrt{3}/6, 0\right).$$

The geometric center of triangle $P_2 P_3 P_4$ is $(0, 0, 0)$ and the geometric center of the tetrahedron is $(0, 0, R\sqrt{6}/12)$.

In order to create a top, let us first invert the tetrahedron shown in Figure 5.1 to the position shown in Figure 5.2, so that

$$(x_1, y_1, z_1) = (0, 0, 0), \qquad (x_2, y_2, z_2) = (0, R\sqrt{3}/3, R\sqrt{6}/3),$$

$$(x_3, y_3, z_3) = \left(\frac{1}{2}R, -R\sqrt{3}/6, R\sqrt{6}/3\right),$$

$$(x_4, y_4, z_4) = \left(-\frac{1}{2}R, -R\sqrt{3}/6, R\sqrt{6}/3\right).$$

The geometric center of triangle $P_2 P_3 P_4$ is now $(0, 0, R\sqrt{6}/3)$, while the geometric center $\bar{P} = (\bar{x}, \bar{y}, \bar{z})$ of the inverted tetrahedron is $(0, 0, R\sqrt{6}/4)$.

5.6 Application – A Spinning Top on a Smooth Horizontal Plane

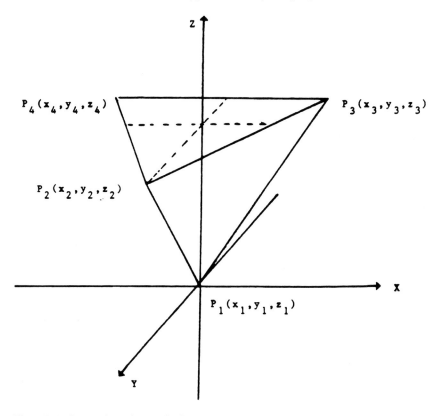

Fig. 5.2 An inverted regular tetrahedron.

Next, let us set P_2, P_3, P_4 in rotation in the plane $z = R\sqrt{6}/3$. As shown in Figure 5.3, we let the velocity of each particle be perpendicular to the line joining that particle to the center of triangle $P_2P_3P_4$. Let P_2, P_3, P_4 have the same speed V. Thus we take the velocities $\vec{v}_i = (v_{ix}, v_{iy}, v_{iz})$, $i = 1, 2, 3, 4$, of P_1, P_2, P_3, P_4 to be

$$\vec{v}_1 = (0,0,0), \quad \vec{v}_2 = (V,0,0),$$
$$\vec{v}_3 = \left(-\frac{1}{2}V, -V\sqrt{3}/2, 0\right), \quad \vec{v}_4 = \left(-\frac{1}{2}V, V\sqrt{3}/2, 0\right).$$

Finally, we want the rotating top to be tilted, initially, relative to the Z axis, so we assume that the line joining P_1 to $\bar{P}$ forms an angle α relative to the Z axis. As shown in Figure 5.4, this will be done by rotating the XZ plane through an angle α. Thus the new positions (x'_i, y'_i, z'_i) and the new velocities

$(v_{ix'}, v_{iy'}, v_{iz'})$ satisfy

$$x'_i = x_i \cos \alpha + z_i \sin \alpha, \quad y'_i = y_i, \quad z'_i = -x_i \sin \alpha + z_i \cos \alpha, \quad (5.38)$$
$$v_{ix'} = v_{ix} \cos \alpha + v_{iz} \sin \alpha, \quad v_{iy'} = v_{iy}, \quad v_{iz'} = -v_{ix} \sin \alpha + v_{iz} \cos \alpha. \quad (5.39)$$

Thus, once the parameters R, V, and α are given, all initial data for a tilted, rotating top are determined.

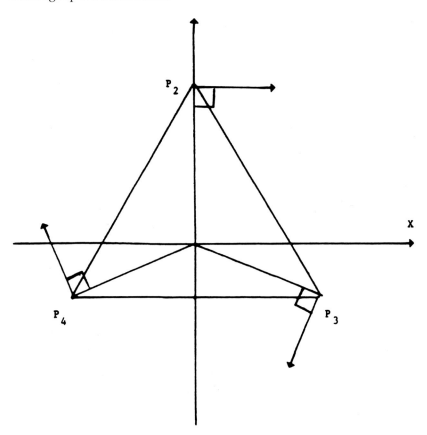

Fig. 5.3 Initial rotation.

The motion of our rotating top is now treated as a four-body problem. At any time t, let P_i, $i = 1, 2, 3, 4$, be located at $\vec{r}_i = (x_i, y_i, z_i)$, have velocity $\vec{v}_i = (\dot{x}_i, \dot{y}_i, \dot{z}_i) = (v_{ix}, v_{iy}, v_{iz})$ and have acceleration $\vec{a} = (\ddot{x}_i, \ddot{y}_i, \ddot{z}_i) = (\dot{v}_{ix}, \dot{v}_{iy}, \dot{v}_{iz})$. Let the mass of each P_i be m_i. For $i \neq j$, let $\vec{r}_{ij}$ be the vector from P_i to P_j and let r_{ij} be the magnitude of $\vec{r}_{ij}$, $i = 1, 2, 3, 4; j = 1, 2, 3, 4; i \neq j$. Let $\phi = \phi(r_{ij})$ be a potential function defined by the pair $P_i, P_j, i \neq j$. Then, for $i = 1, 2, 3, 4$, the Newtonian dynamical equations for the motion of the particles are the

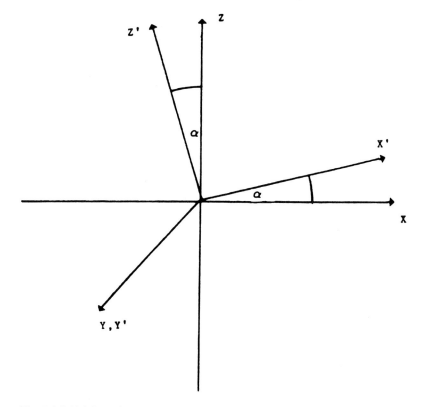

Fig. 5.4 Initial tilt angle α.

second-order differential equations

$$m_i a_{ix} = -\frac{\partial \phi}{\partial r_{ij}} \frac{x_i - x_j}{r_{ij}} - \frac{\partial \phi}{\partial r_{ik}} \frac{x_i - x_k}{r_{ik}} - \frac{\partial \phi}{\partial r_{im}} \frac{x_i - x_m}{r_{im}}$$

$$m_i a_{iy} = -\frac{\partial \phi}{\partial r_{ij}} \frac{y_i - y_j}{r_{ij}} - \frac{\partial \phi}{\partial r_{ik}} \frac{y_i - y_k}{r_{ik}} - \frac{\partial \phi}{\partial r_{im}} \frac{y_i - y_m}{r_{im}}$$

$$m_i a_{iz} = -\frac{\partial \phi}{\partial r_{ij}} \frac{z_i - z_j}{r_{ij}} - \frac{\partial \phi}{\partial r_{ik}} \frac{z_i - z_k}{r_{ik}} - \frac{\partial \phi}{\partial r_{im}} \frac{z_i - z_m}{r_{im}}$$

in which $i = 1$ implies $j = 2$, $k = 3$, $m = 4$; $i = 2$ implies $j = 1$, $k = 3$, $m = 4$; $i = 3$ implies $j = 1$, $k = 2$, $m = 4$; $i = 4$ implies $j = 1$, $k = 2$, $m = 3$. Moreover, we choose $g_1 = 0$, $g_2 = g_3 = g_4 > 0$, so that gravity acts only on P_2, P_3, and P_4.

The above equations are fully conservative.

5 Completely Conservative, Covariant Numerical Methodology

Let us first rewrite the equations as the following equivalent first-order system:

$$\frac{dx_i}{dt} = v_{ix}$$

$$\frac{dy_i}{dt} = v_{iy}$$

$$\frac{dz_i}{dt} = v_{iz}$$

$$m_i \frac{dv_{ix}}{dt} = -\frac{\partial \phi}{\partial r_{ij}} \frac{x_i - x_j}{r_{ij}} - \frac{\partial \phi}{\partial r_{ik}} \frac{x_i - x_k}{r_{ik}} - \frac{\partial \phi}{\partial r_{im}} \frac{x_i - x_m}{r_{im}}$$

$$m_i \frac{dv_{iy}}{dt} = -\frac{\partial \phi}{\partial r_{ij}} \frac{y_i - y_j}{r_{ij}} - \frac{\partial \phi}{\partial r_{ik}} \frac{y_i - y_k}{r_{ik}} - \frac{\partial \phi}{\partial r_{im}} \frac{y_i - y_m}{r_{im}}$$

$$m_i \frac{dv_{iz}}{dt} = -\frac{\partial \phi}{\partial r_{ij}} \frac{z_i - z_j}{r_{ij}} - \frac{\partial \phi}{\partial r_{ik}} \frac{z_i - z_k}{r_{ik}} - \frac{\partial \phi}{\partial r_{im}} \frac{z_i - z_m}{r_{im}} - g_i.$$

To proceed numerically, we next approximate the first-order differential system by the following first-order difference system:

$$\frac{x_{i,n+1} - x_{i,n}}{\Delta t} = \frac{v_{i,n+1,x} + v_{i,n,x}}{2} \tag{5.40}$$

$$\frac{y_{i,n+1} - y_{i,n}}{\Delta t} = \frac{v_{i,n+1,y} + v_{i,n,y}}{2} \tag{5.41}$$

$$\frac{z_{i,n+1} - z_{i,n}}{\Delta t} = \frac{v_{i,n+1,z} + v_{i,n,z}}{2} \tag{5.42}$$

$$m_i \frac{v_{i,n+1,x} - v_{i,n,x}}{\Delta t} = -\frac{\phi(r_{ij,n+1}) - \phi(r_{ij,n})}{r_{ij,n+1} - r_{ij,n}} \cdot \frac{x_{i,n+1} + x_{i,n} - x_{j,n+1} - x_{j,n}}{r_{ij,n+1} + r_{ij,n}}$$
$$- \frac{\phi(r_{ik,n+1}) - \phi(r_{ik,n})}{r_{ik,n+1} - r_{ik,n}} \cdot \frac{x_{i,n+1} + x_{i,n} - x_{k,n+1} - x_{k,n}}{r_{ik,n+1} + r_{ik,n}}$$
$$- \frac{\phi(r_{im,n+1}) - \phi(r_{im,n})}{r_{im,n+1} - r_{im,n}} \cdot \frac{x_{i,n+1} + x_{i,n} - x_{m,n+1} - x_{m,n}}{r_{im,n+1} + r_{im,n}} \tag{5.43}$$

$$m_i \frac{v_{i,n+1,y} - v_{i,n,y}}{\Delta t} = -\frac{\phi(r_{ij,n+1}) - \phi(r_{ij,n})}{r_{ij,n+1} - r_{ij,n}} \cdot \frac{y_{i,n+1} + y_{i,n} - y_{j,n+1} - y_{j,n}}{r_{ij,n+1} + r_{ij,n}}$$
$$- \frac{\phi(r_{ik,n+1}) - \phi(r_{ik,n})}{r_{ik,n+1} - r_{ik,n}} \cdot \frac{y_{i,n+1} + y_{i,n} - y_{k,n+1} - y_{k,n}}{r_{ik,n+1} + r_{ik,n}}$$
$$- \frac{\phi(r_{im,n+1}) - \phi(r_{im,n})}{r_{im,n+1} - r_{im,n}} \cdot \frac{y_{i,n+1} + y_{i,n} - y_{m,n+1} - y_{m,n}}{r_{im,n+1} + r_{im,n}} \tag{5.44}$$

$$m_i \frac{v_{i,n+1,z} - v_{i,n,z}}{\Delta t} = -\frac{\phi(r_{ij,n+1}) - \phi(r_{ij,n})}{r_{ij,n+1} - r_{ij,n}} \cdot \frac{z_{i,n+1} + z_{i,n} - z_{j,n+1} - z_{j,n}}{r_{ij,n+1} + r_{ij,n}}$$
$$- \frac{\phi(r_{ik,n+1}) - \phi(r_{ik,n})}{r_{ik,n+1} - r_{ik,n}} \cdot \frac{z_{i,n+1} + z_{i,n} - z_{k,n+1} - z_{k,n}}{r_{ik,n+1} + r_{ik,n}}$$
$$- \frac{\phi(r_{im,n+1}) - \phi(r_{im,n})}{r_{im,n+1} - r_{im,n}} \cdot \frac{z_{i,n+1} + z_{i,n} - z_{m,n+1} - z_{m,n}}{r_{im,n+1} + r_{im,n}} - g_i. \tag{5.45}$$

Difference equations (5.40)–(5.45) consists of 24 equations in the unknowns $x_{i,n+1}, y_{i,n+1}, z_{i,n+1}, v_{i,n+1,x}, v_{i,n+1,y}, v_{i,n+1,z}$, $i = 1,2,3,4$; and in the knowns $x_{i,n}, y_{i,n}, z_{i,n}, v_{i,n,x}, v_{i,n,y}, v_{i,n,z}$. This system is solved at each time step by Newton's method.

In considering examples, we must first choose a potential function ϕ. We do this in cgs units and in such a fashion that we ensure that the tetrahedron is rigid. To accomplish this, we introduce the following classical, molecular type function:

$$\phi = A\left[-\frac{1}{r_{ij}^3} + \frac{1}{r_{ij}^5}\right], \quad A > 0, \tag{5.46}$$

in which A is sufficiently large to impose rigidity. The choice implies that the magnitude F of the force $\vec{F}$ determined by ϕ satisfies

$$F = A\left[-\frac{3}{r_{ij}^4} + \frac{5}{r_{ij}^6}\right].$$

Thus, $F(\bar{r}) = 0$ provided $\bar{r} = (5/3)^{\frac{1}{2}} \sim 1.290\,994\,449$.

We now choose the tetrahedral edge length R to be

$$R = 1.290\,994\,449.$$

For this value of R, the force between any two of the particles is zero, so that the tetrahedron is physically stable. In the first examples to be described, the parameters A, g, and m_i are scaled for computational convenience to be $A = 10^6$, $g = 0.980$, $m_i \equiv 1$, $i = 1,2,3,4$, unless otherwise specified. The time step Δt is chosen to be $\Delta t = 10^{-5}$. The positions of P_1 and $\bar{P}$ are recorded every 5000 time steps through 11 000 000 time steps. In all the examples, the distance between any two of P_1, P_2, P_3, P_4 is always 1.291. Finally, if at any time any one of z_2, z_3, z_4 is zero, the calculations are stopped and it is concluded that the top has fallen and ceased its motion.

Example 5.1 Set $V = 4$, $\alpha = 15°$. Figure 5.5 shows the cusped path of P_1 in the XY plane using the first 350 points of its trajectory and yields just over one complete cycle. There are 6^+ cycles in the 2200 point trajectory shown in Figure 5.6, which has been enlarged for clarity. The point $\bar{P}$ for the entire 2200 point trajectory oscillates on the line $(0.204\,6144\,211,0,z)$, with $\bar{z}$ rising and falling in the range $0.754 < \bar{z} < 0.764$. Thus, the center point of all the trajectories shown in Figures 5.5 and 5.6 in the XY plane is $(0.204\,614\,4211,0)$.

Example 5.2 Set $V = 8$, $\alpha = 15°$. Figure 5.7 shows the cusped path of P_1 in the XY plane using the first 750 points of its trajectory to yield just over one complete cycle. There are 3^+ cycles in the 2200 point trajectory shown in Figure 5.8, which has been enlarged for clarity. The point $\bar{P}$ for the entire 2200 point trajectory oscillates on the line $(0.204\,614\,421\,1,0,z)$, that is, the same line as in Example 5.1, with $\bar{z}$ rising and falling in the range $0.762 < \bar{z} < 0.764$. Thus, the center point of all the trajectories shown in Figures 5.7 and 5.8 in the XY plane is $(0.204\,614\,421\,1,0)$. Note there are more cusps in Figure 5.7 than in Figure 5.5, but these are smaller, and that the trajectory is becoming more circular.

Example 5.3 Set $V = 16$, $\alpha = 15°$. Figure 5.9 shows a relatively circular trajectory for P_1 in the XY plane using all 2200 points, which yield just over one cycle. The point $\bar{P}$ for the entire trajectory oscillates on the line $(0.204\,614\,421\,1,0,z)$ with $\bar{z}$ rising and falling in the range $0.763 < \bar{z} < 0.764$.

Example 5.4 Set $V = 4$, $\alpha = 30°$. Figure 5.10 shows the cusped path of P_1 in the XY plane using the first 350 points of its trajectory, which yields just over one complete cycle. There are 6^+ cycles in the 2200 point trajectory shown in Figure 5.11, which has been enlarged for clarity. The point $\bar{P}$ for the entire 2200 point trajectory lies on the line $(0.395\,284\,707\,6,0,z)$, with $\bar{z}$ rising and falling in the range $0.652 < \bar{z} < 0.685$.

Example 5.5 Set $V = 8$, $\alpha = 30°$. Figure 5.12 shows the cusped path of P_1 in the XY plane using the first 750 points of its trajectory and yields just over one complete cycle. There are 3^+ cycles in the 2200 point trajectory shown in Figure 5.13, which has been enlarged for clarity. The point $\bar{P}$ for the entire 2200 point trajectory oscillates on the line $(0.395\,284\,707\,6,0,z)$, with $\bar{z}$ rising and falling in the range $0.678 < \bar{z} < 0.685$. Compared to Example 5.4, the number of cusps is increasing, but their size is decreasing.

Example 5.6 Set $V = 16$, $\alpha = 30°$. Figure 5.14 shows a relatively circular path for P_1 in the XY plane using all 2200 trajectory points, which yield just over one cycle. The center of the circle is $(0.395\,284\,707\,6,0)$ The point $\bar{P}$ is always on the line $(0.395\,284\,707\,6,0,z)$, with $\bar{z}$ in the range $0.683 < \bar{z} < 0.685$.).

Example 5.7 Set $V = 4$, $\alpha = 45°$. Figure 5.15 shows the cusped path of P_1 in the XY plane using the first 350 points of its trajectory and yields just over one complete cycle. There are 6^+ cycles in the 2200 point trajectory, which has been enlarged for clarity in Figure 5.16. The point $\bar{P}$ always lies on the line $(0.559\,016\,994\,5,0,z)$, with $\bar{z}$ rising and falling in the range $0.497 < \bar{z} < 0.559$.

5.6 Application – A Spinning Top on a Smooth Horizontal Plane

Example 5.8 Set $V = 8$, $\alpha = 45°$. Figure 5.17 shows the cusped path of P_1 in the XY plane using the first 750 points of its trajectory, which yields just over one complete cycle. There are 3^+ cycles in the 2200 point trajectory shown in Figure 5.18. The point $\bar{P}$ is always on the line $(0.559\,016\,994\,5, 0, z)$, with $\bar{z}$ rising and falling in the range $0.545 < \bar{z} < 0.559$.

Example 5.9 Set $V = 16$, $\alpha = 45°$. Figure 5.19 shows the relatively circular trajectory of P_1 in the XY plane for all 2200 points, which yields just over one complete cycle. The point $\bar{P}$ is always on the line $(0.559\,016\,994\,5, 0, z)$, with $\bar{z}$ in the range $0.555 < \bar{z} < 0.559$.

Example 5.10 Set $V = 1$, $\alpha = 15°$. The top falls to the XY plane.

Examples 5.1–5.9 reveal that each trajectory has a center point, that the trajectories become more circular with increasing V, that the variation in $\bar{z}$ diminishes as V increases, and that the diameter of the trajectory increases with α.

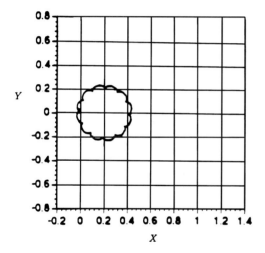

Fig. 5.5 Cusped path of P_1 for $V = 4$, $\alpha = 15°$, showing 350 points.

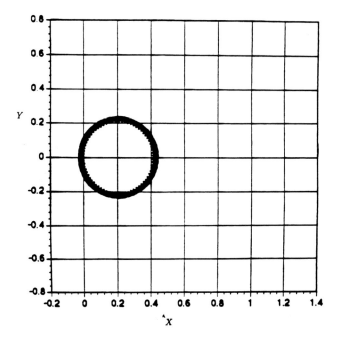

Fig. 5.6 Cusped path of P_1 for $V = 4$, $\alpha = 15°$, showing 2200 points.

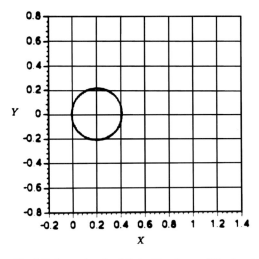

Fig. 5.7 Cusped path of P_1 for $V = 8$, $\alpha = 15°$, showing 750 points.

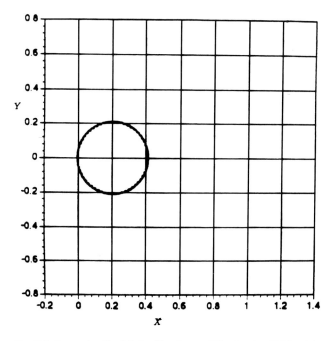

Fig. 5.8 Cusped path of P_1 for $V = 8$, $\alpha = 15°$, showing 2200 points.

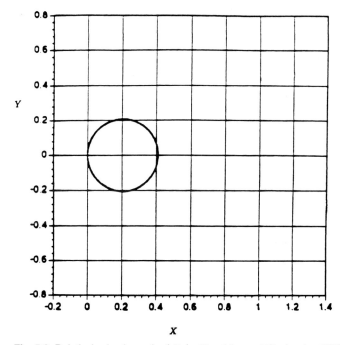

Fig. 5.9 Relatively circular path of P_1 for $V = 16$, $\alpha = 15°$, showing 2200 points.

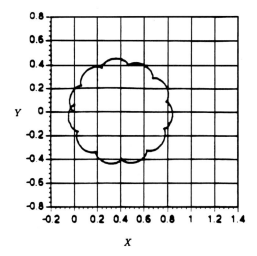

Fig. 5.10 Cusped path of P_1 for $V = 4$, $\alpha = 30°$, showing 350 points.

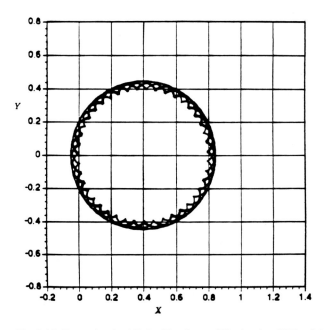

Fig. 5.11 Cusped path of P_1 for $V = 4$, $\alpha = 30°$, showing 2200 points.

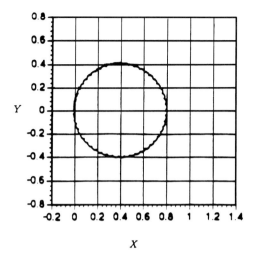

Fig. 5.12 Cusped path of P_1 for $V = 8$, $\alpha = 30°$, showing 750 points.

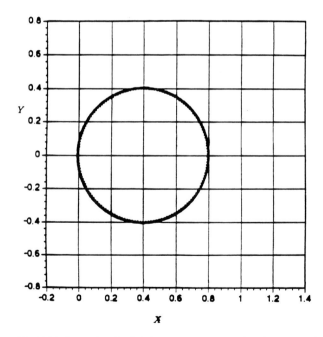

Fig. 5.13 Cusped path of P_1 for $V = 8$, $\alpha = 30°$, showing 2200 points.

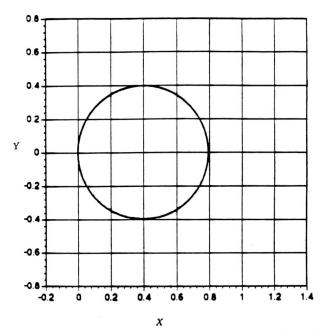

Fig. 5.14 Relatively circular path of P_1 for $V = 16$, $\alpha = 30°$, showing 2200 points.

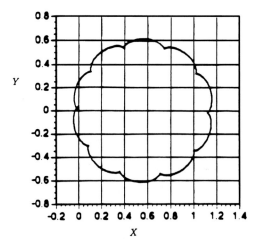

Fig. 5.15 Cusped path of P_1 for $V = 4$, $\alpha = 45°$, showing 350 points.

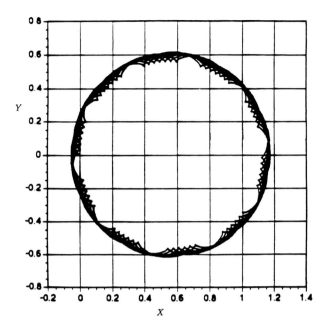

Fig. 5.16 Cusped path of P_1 for $V = 4$, $\alpha = 45°$, showing 2200 points.

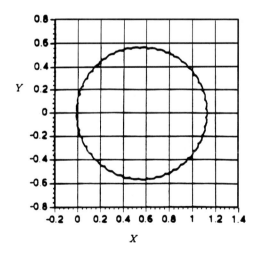

Fig. 5.17 Cusped path of P_1 for $V = 8$, $\alpha = 45°$, showing 750 points.

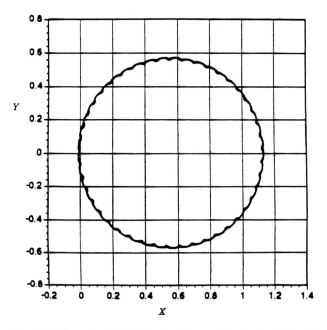

Fig. 5.18 Cusped path of P_1 for $V = 8$, $\alpha = 45°$, showing 2200 points.

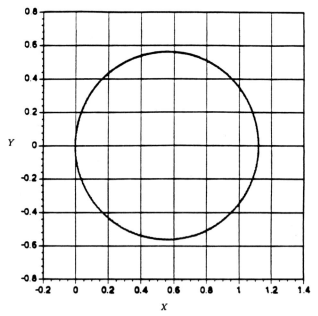

Fig. 5.19 Relatively circular path of P_1 for $V = 16$, $\alpha = 45°$, showing 2200 points.

5.6 Application – A Spinning Top on a Smooth Horizontal Plane

We turn next to the difficult problem of rotating a *nonhomogeneous* top. Note that in earlier sections we used the term *geometric* center throughout rather than *mass* center because of the examples to be considered next.

Example 5.11 Let $V = 4$, $\alpha = 15°$, as in Example 1, but set $m_2 = 0.995$. The resulting 2200 point trajectory for P_1 in the XY plane is shown in Figure 5.20. It shows cusped motion which is similar to that in Figure 5.5 but is in motion to the left. The motion of $\bar{P}$ is fully three-dimensional with the projection of its first 100 points shown in Figure 5.21. Though $\bar{y}$ and $\bar{z}$ show only small variations, it is $\bar{x}$ that shows significant motion.

Example 5.12 Let $V = 4$, $\alpha = 30°$, as in Example 5.4, but set $m_2 = 1.005$. The results are similar to those in Example 5.11 and are shown in Figures 5.22 and 5.23. However, this time the motion of P_1 in the XY plane is to the right.

Example 5.13 Let $V = 8$, $\alpha = 30°$, as in Example 5.5, but set $m_2 = 1.005$ and $m_3 = 0.95$. The entire motion for P_1 is shown in Figure 5.24 and the projection for $\bar{P}$, but for only the first 100 points, is shown in Figure 5.25. Figure 5.24 shows complex looping motion up and to the right.

Example 5.14 Let $V = 4$, $\alpha = 45°$, $m_2 = 0.995$, $m_3 = 0.985$. The resulting motion of P_1 for the entire 2200 points is shown in Figure 5.26 and the projection for $\bar{P}$, but for only the first 100 points, is shown in Figure 5.27. The motion of P_1 is up and to the right, but similar to that of Figure 5.22.

With regard to Examples 5.11–5.14 and the corresponding graphs in Figures 5.21, 5.23, 5.25, and 5.27, it is worth noting that these graphs, with only 100 points, characterize completely the graphs with 2000 points.

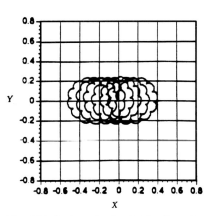

Fig. 5.20 Projected path of P_1 for $V = 4$, $\alpha = 15°$, showing 2200 points.

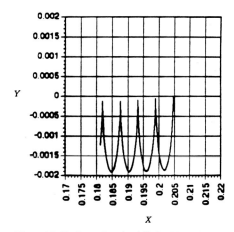

Fig. 5.21 Projected path of $\bar{P}_1$ for $V = 4$, $\alpha = 15°$, showing 100 points.

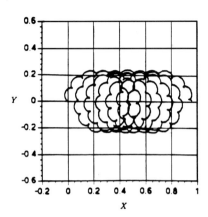

Fig. 5.22 Projected path of P_1 for $V = 4$, $\alpha = 30°$, showing 2200 points.

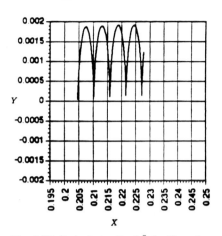

Fig. 5.23 Projected path of $\bar{P}_1$ for $V = 4$, $\alpha = 30°$, showing 100 points.

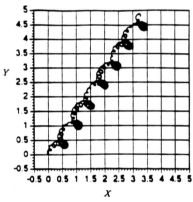

Fig. 5.24 Projected path of P_1 for $V = 8$, $\alpha = 30°$, showing 2200 points.

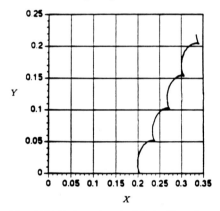

Fig. 5.25 Projected path of $\bar{P}_1$ for $V = 8$, $\alpha = 30°$, showing 100 points.

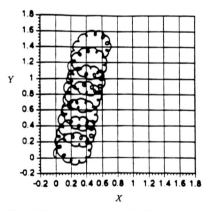

Fig. 5.26 Projected path of P_1 for $V = 4$, $\alpha = 45°$, showing 2200 points.

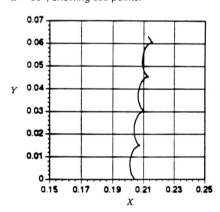

Fig. 5.27 Projected path of $\bar{P}_1$ for $V = 4$, $\alpha = 45°$, showing 100 points.

Finally, we give in generic form a computer program for simulating top rotation. The parameters are those of Example 5.1 in Section 5.6.

Algorithm 7 Program TOP

Step 1. Set a time step equal to 10^{-5}.
Step 2. Set a Newtonian convergence tolerance equal to $5(10)^{-12}$.
Step 3. Let the four particles $P(I)$, $I = 1, 2, 3, 4$, each have unit mass.
Step 4. For $\alpha = 15°$, $V = 4$, $R = 1.290\,994\,449$, determine the initial positions and velocities of each $P(I)$ by (5.38), (5.39).
Step 5. Fix $A = 10^5$ in (5.46)
Step 6. Beginning with $n = 0$, and proceeding in sequence, determine for each fixed n the new positions $\vec{r}_{n+1}(I)$ and new velocities $\vec{v}_{n+1}(I)$ in terms of the known $\vec{r}_n(I)$ and $\vec{v}_n(I)$, $I = 1, 2, 3, 4$. This is done by solving the 24 equations (5.40)–(5.45) by Newton's method. However, the equations should be simplified first by removing the singularity $r_{n+1} - r_n$ in each of (5.43)–(5.45) by actually carrying out each of the divisions
$$\frac{\phi_{n+1} - \phi_n}{r_{n+1} - r_n}.$$
Step 7. Stop the calculations when $n = 2500$.

For a direct extension of the top model given in this section, see Greenspan (1998).

5.7 Application – Calogero and Toda Hamiltonian Systems

A Calogero system (Calogero (1975)) is a system of n particles on a line with Hamiltonian

$$H = \frac{1}{2} \sum_{i=1}^{n} p_i^2 + \sum_{i<j}^{n-1} \frac{1}{(q_i - q_j)^2}. \tag{5.47}$$

A Toda lattice (Toda (1967)) is a system of n particles on a line with Hamiltonianindex Toda system

$$H = \frac{1}{2} \sum_{i=1}^{n} p_i^2 + \sum_{i=1}^{n-1} \exp(q_i - q_{i+1}). \tag{5.48}$$

Both equations (5.47) and (5.48) have received extensive theoretical study (see, for example, Marsden (1981), Moser (1980), and the numerous references contained therein). For each

$$\sum_{i=1}^{n} p_i \tag{5.49}$$

is a system invariant.

For clarity and intuition, let us begin with a two-particle Calogero system whose Hamiltonian is

$$H = \frac{1}{2}\sum_{i=1}^{2} p_i^2 + \frac{1}{(q_1-q_2)^2}, \quad q_1 \neq q_2 \tag{5.50}$$

Let $\Delta t > 0$ and $t_k = k\Delta t$, $k = 0, 1, 2, 3, \ldots$. Denote p_1, p_2, q_1, q_2 at time t_k by $p_{1,k}, p_{2,k}, q_{1,k}, q_{2,k}$, respectively. For $i = 1, 2$ and $k = 0, 1, 2, 3, \ldots$, define

$$\frac{p_{i,k+1} + p_{i,k}}{2} = \frac{q_{i,k+1} - q_{i,k}}{\Delta t} \tag{5.51}$$

$$\frac{p_{i,k+1} - p_{i,k}}{\Delta t} = F_{i,k}, \tag{5.52}$$

where

$$F_{1,k} = \frac{q_{1,k+1} + q_{1,k} - q_{2,k+1} - q_{2,k}}{(q_{1,k} - q_{2,k})^2 (q_{1,k+1} - q_{2,k+1})^2} \tag{5.53}$$

$$F_{2,k} = -F_{1,k}. \tag{5.54}$$

Then we have the following theorems.

Theorem 5.7 *Given $p_{1,0}, p_{2,0}, q_{1,0}, q_{2,0}$, then equations (5.52)–(5.54) imply the invariance of $p_{1,k} + p_{2,k}$, that is,*

$$p_{1,k} + p_{2,k} = p_{1,0} + p_{2,0}, \quad k = 1, 2, 3, \ldots. \tag{5.55}$$

Proof. From (5.52)–(5.54),

$$p_{1,k+1} = p_{1,k} + (\Delta t) F_{1,k}$$
$$p_{2,k+1} = p_{2,k} + (\Delta t) F_{2,k}.$$

Hence, for $k = 0, 1, 2, 3, \ldots$,

$$p_{1,k+1} + p_{2,k+1} = p_{1,k} + p_{2,k}$$

which implies (5.55). □

Theorem 5.8 *Difference formulas (5.51)–(5.54) imply the invariance of Hamiltonian (5.50) for given $p_{1,0}, p_{2,0}, q_{1,0}, q_{2,0}$, that is, for $k = 1, 2, 3, \ldots$,*

$$\frac{1}{2}\left[(p_{1,k})^2 + (p_{2,k})^2\right] + \frac{1}{(q_{1,k} - q_{2,k})^2}$$
$$= \frac{1}{2}\left[(p_{1,0})^2 + (p_{2,0})^2\right] + \frac{1}{(q_{1,0} - q_{2,0})^2} \tag{5.56}$$

Proof. Let

$$W_n = \sum_{k=0}^{n-1} \left[(q_{1,k+1} - q_{1,k})F_{1,k} + (q_{2,k+1} - q_{2,k})F_{2,k} \right]. \tag{5.57}$$

Then, from (5.51) and (5.52),

$$W_n = \sum_{k=0}^{n-1} \left[(q_{1,k+1} - q_{1,k}) \frac{p_{1,k+1} - p_{1,k}}{\Delta t} + (q_{2,k+1} - q_{2,k}) \frac{p_{2,k+1} - p_{2,k}}{\Delta t} \right]$$

$$= \frac{1}{2} \sum_{k=0}^{n-1} \left[(p_{1,k+1})^2 - (p_{1,k})^2 + (p_{2,k+1})^2 - (p_{2,k})^2 \right]$$

so that

$$W_n = \frac{1}{2} \left[(p_{1,n})^2 + (p_{2,n})^2 \right] - \frac{1}{2} \left[(p_{1,0})^2 + (p_{2,0})^2 \right] \tag{5.58}$$

However, (5.53), (5.54) and (5.57) imply

$$W_n = \sum_{k=0}^{n-1} \left[(q_{1,k+1} - q_{1,k}) \frac{q_{1,k+1} + q_{1,k} - q_{2,k+1} - q_{2k}}{(q_{1,k} - q_{2,k})^2 (q_{1,k+1} - q_{2,k+1})^2} \right.$$
$$\left. - (q_{2,k+1} - q_{2,k}) \frac{q_{1,k+1} + q_{1,k} - q_{2,k+1} - q_{2k}}{(q_{1,k} - q_{2,k})^2 (q_{1,k+1} - q_{2,k+1})^2} \right]$$

$$= \sum_{k=0}^{n-1} \frac{(q_{1,k+1} - q_{2,k+1})^2 - (q_{1,k} - q_{2,k})^2}{(q_{1,k} - q_{2,k})^2 (q_{1,k+1} - q_{2,k+1})^2}$$

$$= \sum_{k=0}^{n-1} \left[\frac{1}{(q_{1,k} - q_{2,k})^2} - \frac{1}{(q_{1,k+1} - q_{2,k+1})^2} \right]$$

so that

$$W_n = \frac{1}{(q_{1,0} - q_{2,0})^2} - \frac{1}{(q_{1,n} - q_{2,n})^2}. \tag{5.59}$$

Elimination of W_n between (5.58) and (5.59) and setting $n = k$ then yields (5.56), and the theorem is proved. □

The extension to systems of N particles follows directly from formulation (5.51)–(5.54), but with $i = 1, 2, 3, \ldots, N$.

For a Toda lattice, formulas (5.51)–(5.54) need be modified only slightly, that is, (5.53) needs to be changed and this is done as follows:

$$F_{1,k} = \begin{cases} -\frac{\exp(q_{1,k+1} - q_{2,k+1}) - \exp(q_{1,k} - q_{2,k})}{(q_{1,k+1} - q_{2,k+1}) - (q_{1,k} - q_{2,k})}, & (q_{1,k+1} - q_{2,k+1}) - (q_{1,k} - q_{2,k}) \neq 0 \\ -\exp(q_{1,k} - q_{2,k}), & (q_{1,k+1} - q_{2,k+1}) - (q_{1,k} - q_{2,k}) = 0. \end{cases} \tag{5.60}$$

Theorem 5.9 Given $p_{1,0}, p_{2,0}, q_{1,0}, q_{2,0}$, then (5.51), (5.52), (5.54) and (5.60) imply (5.55).

Proof. The proof is essentially identical to that of Theorem 5.7. □

Theorem 5.10 Under the assumptions of Theorem 5.9, it follows for $k = 1, 2, \ldots$, that

$$\frac{1}{2}\left[(p_{1,k})^2 + (p_{2,k})^2\right] + \exp(q_{1,k} - q_{2,k})$$
$$= \frac{1}{2}\left[(p_{1,0})^2 + (p_{2,0})^2\right] + \exp(q_{1,0} - q_{2,0}). \quad (5.61)$$

Proof. Consider first the case $(q_{1,k+1} - q_{2,k+1}) - (q_{1,k} - q_{2,k}) \neq 0$. Recall also (5.57), that is,

$$W_n = \sum_{k=0}^{n-1}\left[(q_{1,k+1} - q_{1,k})F_{1,k} + (q_{2,k+1} - q_{2,k})F_{2,k}\right]$$

Then, (5.58), that is,

$$W_n = \frac{1}{2}\left[(p_{1,n})^2 + (p_{2,n})^2\right] - \frac{1}{2}\left[(p_{1,0})^2 + (p_{2,0})^2\right]$$

is again valid by the same argument used to derive (5.58).

Next, (5.51), (5.52), (5.54), (5.57) and (5.60) imply

$$W_n = \sum_{k=0}^{n-1}\left\{(q_{1,k+1} - q_{1,k})\left[-\frac{\exp(q_{1,k+1} - q_{2,k+1}) - \exp(q_{1,k} - q_{2k})}{(q_{1,k+1} - q_{2,k+1}) - (q_{1,k} - q_{2,k})}\right]\right.$$
$$\left. + (q_{2,k+1} - q_{2,k})\left[\frac{\exp(q_{1,k+1} - q_{2,k+1}) - \exp(q_{1,k} - q_{2,k})}{(q_{1,k+1} - q_{2,k+1}) - (q_{1,k} - q_{2,k})}\right]\right\}$$

so that

$$W_n = \sum_{k=0}^{n-1}\left\{-\left[\exp(q_{1,k+1} - q_{2,k+1}) - \exp(q_{1,k} - q_{2k})\right]\right\}. \quad (5.62)$$

Hence,

$$W_n = \exp(q_{1,0} - q_{2,0}) - \exp(q_{1,n} - q_{2,n}) \quad (5.63)$$

Finally, elimination of W_n between (5.58) and (5.63) and setting $n = k$ yields (5.61).

In the second case, when $(q_{1,k+1} - q_{2,k+1}) - (q_{1,k} - q_{2,k}) = 0$, the corresponding summation term in (5.57) becomes simply

$$[(q_{1,k+1} - q_{1,k}) - (q_{2,k+1} - q_{2,k})][-\exp(q_{1,k} - q_{2,k})],$$

which is zero, and the theorem continues to be valid. □

5.7 Application – Calogero and Toda Hamiltonian Systems

Let us indicate finally that the difference formulations just developed are practical in that they can be implemented readily for computation, as the following two examples illustrate.

Consider the Calogero difference equations (5.51)–(5.54) with initial data $q_{1,0} = 1$, $q_{2,0} = -1$, $p_{1,0} = 1$, $p_{2,0} = -1$. Then the iteration formulas for solving this system at t_{k+1} in terms of data at t_k are

$$q_{1,k+1}^{(n+1)} = q_{1,k} + \frac{\Delta t}{2}\left[p_{1,k+1}^{(n)} + p_{1,k}\right] \tag{5.64}$$

$$q_{2,k+1}^{(n+1)} = q_{2,k} + \frac{\Delta t}{2}\left[p_{2,k+1}^{(n)} + p_{2,k}\right] \tag{5.65}$$

$$p_{1,k+1}^{(n+1)} = p_{1,k} + (\Delta t)\left[\frac{q_{1,k+1}^{(n+1)} + q_{1,k} - q_{2,k+1}^{(n+1)} - q_{2,k}}{(q_{1,k} - q_{2,k})^2 (q_{1,k+1}^{(n+1)} - q_{2,k+1}^{(n+1)})^2}\right] \tag{5.66}$$

$$p_{2,k+1}^{(n+1)} = p_{2,k} - (\Delta t)\left[\frac{q_{1,k+1}^{(n+1)} + q_{1,k} - q_{2,k+1}^{(n+1)} - q_{2,k}}{(q_{1,k} - q_{2,k})^2 (q_{1,k+1}^{(n+1)} - q_{2,k+1}^{(n+1)})^2}\right] \tag{5.67}$$

Calculation for 500 000 time steps on a VAX 8700 with $\Delta t = 0.0001$ yields in 35 s the results recorded every 50 000 time steps in Table 5.1. The table shows clearly that both the Hamiltonian and $p_1 + p_2$ are conserved throughout. In addition, it shows an increasingly repulsive effect which the particles exert on each other.

Table 5.1 Computational results for Calogero system.

k	H	q_1	q_2	p_1	p_2
1	1.25000	1.00000	−1.00000	1.00000	−1.00000
50000	1.25000	6.49989	−6.49989	1.11539	−1.11539
100000	1.25000	12.0829	−12.0829	1.11727	−1.11727
150000	1.25000	17.6705	−17.6705	1.11767	−1.11767
200000	1.25000	23.2593	−23.2593	1.11783	−1.11783
250000	1.25000	28.8486	−28.8486	1.11790	−1.11790
300000	1.25000	34.4382	−34.4382	1.11794	−1.11794
350000	1.25000	40.0280	−40.0280	1.11797	−1.11797
400000	1.25000	45.6179	−45.6179	1.11798	−1.11798
450000	1.25000	51.2078	−51.2078	1.11799	−1.11799
500000	1.25000	56.7978	−56.7978	1.11800	−1.11800

Using formulas entirely analogous to (5.64)–(5.67), but which incorporate (5.60) for the Toda lattice, calculations for 250 000 steps with $\Delta t = 0.000\,001$ and initial data $q_{1,0} = 1$, $q_{2,0} = -1$, $p_{1,0} = 10$, $p_{2,0} = -10$ yield in 90 s the results in Table 5.2. The second part of formula (5.60) is essential numerically at the turning point which occurs between $k = 190\,000$ and $k = 200\,000$. The table indicates clearly, again, the invariance of both H and $p_1 + p_2$.

Table 5.2 Computational results for Toda system.

k	H	q_1	q_2	p_1	p_2
1	107.39	1.00000	−1.00000	10.00000	−10.00000
50000	107.39	1.48679	−1.48679	9.37162	−9.37162
100000	107.39	1.92153	−1.92153	7.79240	−7.79240
150000	107.39	2.23627	−2.23627	4.45092	−4.45092
200000	107.39	2.33650	−2.33650	−0.60920	0.60920
250000	107.39	2.18026	−2.18026	−5.39371	5.39371

5.8
Remarks

It has been known for some time that, if one wishes to conserve only energy and linear momentum, formulas of higher order than those presented in Section 5.3 are available (LaBudde and Greenspan (1976)). However, research to date has never produced formulas except those of Section 5.3 which, in addition, conserve angular momentum.

There has also been recent research toward a generalization of the methodology of this chapter using amorphous, cellular automata models (Rauch (2003)).

Finally, observe that the discussion in Section 5.6 extends to gyroscopic motion. For the conservative motion of the dodecahedral gyroscope shown in Figure 5.28, see Greenspan (2002).

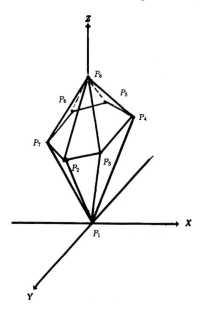

Fig. 5.28 The dodecahedron.

5.9 Exercises

5.1 For the classical gravitational three-body problem, $\phi = -Gm_1m_2/r_{ij}$, in which G is a universal constant. For this choice of ϕ, what are explicit forms of equations (5.1)–(5.3)?

5.2 Prove that equation (5.6) is valid.

5.3 Prove that equation (5.9) is valid.

5.4 Provide the details of Theorem 5.3.

5.5 Complete the proof of Lemma 5.1.

5.6 Provide the details of Theorem 5.4.

5.7 Provide the details of Theorem 5.5.

5.8 Prove that equations (5.43)–(5.45) are valid.

5.9 Extend Theorem 5.1 to the general N-body problem.

5.10 Extend Theorem 5.2 to the general N-body problem.

5.11 Extend Theorem 5.3 to the general N-body problem.

5.12 Extend Theorem 5.4 to three dimensions.

5.13 Extend Theorem 5.5 to three dimensions.

5.14 Extend Theorem 5.6 to three dimensions.

5.15 Show that the difference equation approximation

$$\frac{x_{k+1} - x_k}{\Delta t} = \frac{v_{k+1} + v_k}{2}$$

$$\frac{v_{k+1} - v_k}{\Delta t} = -\frac{K^2}{m} \frac{x_{k+1} + x_k}{2}$$

for the classical harmonic oscillator

$$m\ddot{x} + K^2 x = 0$$

conserves the same energy as the classical oscillator.

6
Instability

6.1
Introduction

The concepts of stability and its negative, instability, have a long history in the development of analytical and qualitative theory of ordinary differential equations (Cesari (1959)). Yet, no single definition has proved to be adequate for all occasions, each of which is reasonable in certain circumstances. The same situation exists in defining stability of difference equations and of numerical solutions of difference equations (Dahlquist and Bjorck (1974)). In this chapter we wish to avoid the multiplicity of mathematical definitions and consider a computer definition only. Our motivation lies in the fact that once a computation yields overflow, that computation ceases.

6.2
Instability Analysis

In solving initial value problems on a computer, the occurrence of overflow is a relatively common and disturbing event. Any such computation will be called *unstable*. This definition is the computer analog of a classical mathematical definition in which a solution of a differential equation is called unstable if it is unbounded on $0 < x < \infty$. Our definition simply considers "unbounded" as being greater than the largest number in one's computer. A computation which is *not* unstable is said to be *stable*.

There are several steps one can take to try to remedy an unstable computation, although none may work. These are as follows. First, *check the program*. This can be done by explaining it to another person or by testing it with a problem for which the solution is known. In explaining it to a second person, one must read one's program word by word, rather than in blocks of words. One

Numerical Solution of Ordinary Differential Equations for Classical, Relativistic and Nano Systems. Donald Greenspan
Copyright © 2006 WILEY-VCH Verlag GmbH & Co. KGaA, Weinheim
ISBN: 3-527-40610-7

can always create a problem with a known solution directly from the problem under consideration. Suppose, for example, one is solving the following initial value problem numerically and instability results:

$$y'' + xy + xy^2 = 1, \quad y(0) = y'(0) = 1.$$

Consider then the equation

$$y'' + xy + xy^2 = 1 + f(x)$$

and take as a solution of this equation $y = x^2$. Then by direct substitution into the latter equation, one finds that

$$2 + x^3 + x^5 - 1 \equiv f(x).$$

One then tests the program by applying it to the problem

$$y'' + xy + xy^2 = 2 + x^3 + x^5, \quad y(0) = 0, \quad y'(0) = 0,$$

the solution of which is $y = x^2$.

If the instability persists, then one next *checks the computer*. Overheating can cause erratic computer performance and miswiring has been known to occur. The most practical check is to run the same program on a second computer.

If instability still persists, then one can attempt a mathematical analysis, when possible. Such analyses are given in the following examples.

Example 6.1 Consider the initial value problem

$$y' = -100y \tag{6.1}$$

$$y(0) = 1. \tag{6.2}$$

The exact solution is $y(x) = e^{-100x}$, which converges to zero as x goes to infinity and is always positive. Suppose one solves (6.1), (6.2) numerically with Euler's method using $h = 0.1$. Then

$$\frac{y_{i+1} - y_i}{0.1} = -100y_i, \quad y_0 = 1; \, i = 0, 1, 2, 3, \ldots,$$

or,

$$y_{i+1} = -9y_i, \quad y_0 = 1; \, i = 0, 1, 2, 3, \ldots.$$

Then,

$$y_1 = -9$$
$$y_2 = (-9)^2$$
$$y_3 = (-9)^3$$
$$y_4 = (-9)^4$$
$$\vdots$$

and so forth, which yields overflow quickly. To analyze this instability, we redo the calculation but *do not* fix h. Hence,

$$\frac{y_{i+1} - y_i}{h} = -100 y_i, \quad y_0 = 1; i = 0, 1, 2, 3, \ldots,$$

so that

$$y_{i+1} = (1 - 100h) y_i, \quad y_0 = 1; i = 0, 1, 2, 3, \ldots.$$

Thus,

$$\begin{aligned} y_1 &= (1 - 100h) y_0 = (1 - 100h) \\ y_2 &= (1 - 100h) y_1 = (1 - 100h)^2 \\ y_3 &= (1 - 100h)^3 \\ y_4 &= (1 - 100h)^4 \\ &\vdots \end{aligned}$$

from which it follows that

$$y_i = (1 - 100h)^i, \quad i = 0, 1, 2, 3, \ldots.$$

For stability, one must have

$$|1 - 100h| \leq 1,$$

or,

$$-1 \leq 1 - 100h \leq 1,$$

or

$$-2 \leq -100h \leq 0.$$

Thus, one requires

$$100h \leq 2$$

or

$$h \leq 0.02. \tag{6.3}$$

The constraint (6.3) is called a stability condition on the grid size. The choice $h = 0.1$ violated this condition. Moreover, when the strict inequality in (6.3) is satisfied, the numerical solution converges to zero as x_i goes to infinity.

In the above example, the equation (6.1) was linear. If the equation were nonlinear, one often linearizes it and develops an approximate stability condition. Sometimes one can develop energy inequalities for a stability analysis (Cryer (1969); Richtmyer and Morton (1967)), as shown in a later example, but these may be very difficult to obtain. In any case, the first rule of thumb to be derived is that if instability occurs, *decrease the grid size* and repeat the calculation. This rule works so well that it often solves the problem. In some cases, however, it may not. We consider a related example next.

6 Instability

Example 6.2 Suppose one is given the following recursion formula and additional conditions:

$$y_{i+1} = -\frac{3}{2}y_{i+1} + y_i \quad y_0 = \frac{1}{2}, \quad y_1 = \frac{1}{4}, \quad i = 0, 1, 2, 3, \ldots . \tag{6.4}$$

Then

$$y_2 = -\frac{3}{2}y_1 + y_0 = \frac{1}{8}$$
$$y_3 = -\frac{3}{2}y_2 + y_1 = \frac{1}{16}$$
$$y_4 = -\frac{3}{2}y_3 + y_2 = \frac{1}{32}$$
$$\vdots$$

from which it follows that

$$y_i = \left(\frac{1}{2}\right)^{i+1}, \quad i = 0, 1, 2, 3, \ldots$$

which converges to zero as i goes to infinity. Suppose, however, as in all computers, one has to round. For simplicity, we will round all y values to one decimal place. Rounding to a higher number of decimal places merely delays the onset of the phenomenon to be shown next. Hence, recalculating with rounding, we find first that (6.4) becomes

$$y_{i+2} = -(1.5)y_{i+1} + y_i, \quad y_0 = 0.5, \quad y_1 = 0.3. \tag{6.5}$$

Thus,

$$y_2 = -(1.5)y_1 + y_0 = 0.1$$
$$y_3 = -(1.5)y_2 + y_1 = 0.2$$
$$y_4 = -(1.5)y_3 + y_2 = -0.2$$
$$y_5 = -(1.5)y_4 + y_3 = 0.5$$
$$y_6 = -(1.5)y_5 + y_4 = -1.0$$
$$y_7 = -(1.5)y_6 + y_5 = 2.0$$
$$y_8 = -(1.5)y_7 + y_6 = -4.0$$
$$y_9 = 8.0$$
$$y_{10} = -16.0$$
$$\vdots$$

$$y_k = -(-2)^{k-6}, \quad k \geq 6,$$

so that y_i becomes unbounded with increasing i. Let us show how to analyze this instability.

First, let us rewrite (6.4) as

$$y_{i+2} + 3y_{i+1} - 2y_i = 0. \tag{6.6}$$

Equation (6.6) is called a second-order linear difference equation with constant coefficients. In analogy with second-order linear differential equations with constant coefficients, it can be solved in general as follows. Set

$$y_i = \lambda^i, \quad \lambda \text{ being a nonzero constant.}$$

Then,

$$2\lambda^{i+2} + 3\lambda^{i+1} - 2\lambda^i = 0$$

or,

$$2\lambda^2 + 3\lambda - 2 = 0.$$

Thus,

$$(2\lambda - 1)(\lambda + 2) = 0.$$

and $\lambda = \frac{1}{2}, \lambda = -2$. Each of

$$y_i = \left(\frac{1}{2}\right)^i, \quad y_i = (-2)^i$$

is a solution of (6.6) and the general solution is

$$y_i = c_1 \left(\frac{1}{2}\right)^i + c_2(-2)^i, \quad i = 0, 1, 2, 3, \ldots \tag{6.7}$$

in which c_1, c_2 are arbitrary constants. For our first calculation above, we had $y_0 = \frac{1}{2}, y_1 = \frac{1}{4}$. Thus from (6.7), for $i = 0$ and $i = 1$,

$$c_1 + c_2 = \frac{1}{2}$$

$$c_1 \left(\frac{1}{2}\right) + c_2(-2) = \frac{1}{4}$$

which imply $c_1 = \frac{1}{2}, c_2 = 0$. Thus,

$$y_i = \left(\frac{1}{2}\right)^{i+1},$$

which is the same as the result we got when the calculations were exact. But, when including rounding, we had $y_0 = 0.5, y_1 = 0.3$. For $i = 0$ and $i = 1$, then, from (6.7),

$$c_1 + c_2 = 0.5$$

$$c_1 \left(\frac{1}{2}\right) + c_2(-2) = 0.3.$$

The solution to this system is $c_1 = \frac{13}{25}$, $c_2 = -\frac{1}{50}$, and the general solution is now

$$y_i = \frac{13}{25}\left(\frac{1}{2}\right)^i - \frac{1}{50}(-2)^i, \quad i = 0, 1, 2, 3, \ldots. \tag{6.8}$$

However, in (6.8), the term $(-2)^i$ becomes unbounded as i becomes large, which is why the instability results.

Note that the difference equation (6.6) can be regarded as an approximation for a first-order differential equation. Consider, in fact, equation (6.1), that is,

$$y' = -100y. \tag{6.9}$$

Fix $h > 0$ and consider the grid points $x_i = ih$, $i = 0, 1, 2, 3, \ldots$. At x_{i+1}, $i = 0, 1, 2, 3, \ldots$, approximation of y' by a central difference approximation yields in (6.1)

$$\frac{y_{i+2} - y_i}{2h} = -100 y_{i+1}$$

or, equivalently,

$$y_{i+1} + 2(100) h y_{i+1} - y_i = 0. \tag{6.10}$$

For $h = 3/400$, (6.10) is equivalent to (6.6).

Note, finally, that the difference equation (6.10) leads to unstable computation for any positive h. In fact, the general solution of (6.10) is

$$y_i = c_1 (\lambda_1)^i + c_2 (\lambda_2)^i,$$

where λ_1, λ_2 are given by

$$\lambda_{1,2} = -100h \pm \sqrt{(100h)^2 + 1}.$$

Thus, no matter how small we choose h, since

$$\left| -100h - \sqrt{(100h)^2 + 1} \right| > 1$$

one of $|\lambda_1|, |\lambda_2|$ is always greater than unity and the resulting calculations will be unstable for any nonzero choice of c_1 and c_2.

From the above example follows the second rule of thumb. If decreasing the grid size always continues to yield instability, then *change the difference equation being used*. The way to do this is to always choose a formula of the form

$$a y_{i+2} + b y_{i+1} + c y_i = 0, \quad a \neq 0,$$

such that the two roots λ_1, λ_2 of

$$a \lambda^2 + b \lambda + c = 0$$

satisfy $|\lambda| \leq 1$. The reason is that if this condition is valid then all particular solutions of the general solution

$$y_i = c_1(\lambda_1)^i + c_2(\lambda_2)^i$$

are bounded for all i.

Let us turn now to a nonlinear problem. For "highly" nonlinear problems, mathematical analysis of instability is rarely available. The problem we will examine is a pendulum problem which is classified often as a "mildly" nonlinear problem. However, because our definition of instability is machine dependent, in this example we will concentrate only on mathematical stability, that is, we will show that the exact sequence of values generated by the method to be used is bounded.

Example 6.3 Consider the initial value problem

$$\ddot{x} + 1.6\dot{x} + 32 \sin x = 0, \quad x_0 = \pi/4, \quad v_0 = 0. \tag{6.11}$$

Let us solve this problem numerically using the recursion formulas

$$\frac{x_{k+1} - x_k}{h} = \frac{v_{k+1} + v_k}{2} \tag{6.12}$$

$$\frac{v_{k+1} - v_k}{h} = a_k \tag{6.13}$$

$$a_k = -1.6v_k - 32 \sin x_k \tag{6.14}$$

or, equivalently, by

$$x_{k+1} - x_k = \frac{1}{2}h(v_{k+1} + v_k) \tag{6.15}$$

$$v_{k+1} - v_k = -h(1.6v_k + 32 \sin x_k). \tag{6.16}$$

Our procedure will be to manipulate (6.15) and (6.16) to obtain energy inequalities which will then yield the desired stability condition. Thus, multiplication of (6.16) by $(v_{k+1} + v_k)/2$ implies

$$\frac{1}{2}(v_{k+1})^2 - \frac{1}{2}(v_k)^2 = -\frac{1}{2}h(v_{k+1} + v_k)(1.6)v_k - \frac{1}{2}h(32 \sin x_k)(v_{k+1} + v_k),$$

or, from (6.12),

$$\frac{1}{2}(v_{k+1})^2 - \frac{1}{2}(v_k)^2 = -\frac{1}{2}(1.6)h(v_{k+1} + v_k)v_k - (32 \sin x_k)(x_{k+1} - x_k). \tag{6.17}$$

Observe that for mass $m = 1$, the left side of (6.17) is the difference of two consecutive kinetic energies. It is necessary then to introduce the potential energies and the damping. Note, also, that the damping coefficient 1.6 will not be simplified for reasons to be discussed later.

6 Instability

Since, by finite Taylor expansion,

$$\cos x_{k+1} = \cos x_k - (x_{k+1} - x_k)\sin x_k - \frac{1}{2}(x_{k+1} - x_k)^2 \cos \xi,$$

then

$$-(32\cos x_{k+1} - 32\cos x_k)$$
$$= (x_{k+1} - x_k)(32)\sin x_k + \frac{1}{2}(x_{k+1} - x_k)^2(32\cos \xi),$$

so that

$$-(32\cos x_{k+1} - 32\cos x_k) \leq (x_{k+1} - x_k)(32)\sin x_k + 16(x_{k+1} - x_k)^2. \quad (6.18)$$

Addition of (6.17) and (6.18) implies

$$\frac{1}{2}(v_{k+1})^2 - \frac{1}{2}(v_k)^2 - (32\cos x_{k+1} - 32\cos x_k)$$
$$\leq -\frac{1.6h}{2}(v_{k+1} + v_k)v_k + 16(x_{k+1} - x_k)^2,$$

or, from (6.12),

$$\frac{1}{2}(v_{k+1})^2 - \frac{1}{2}(v_k)^2 - (32\cos x_{k+1} - 32\cos x_k)$$
$$\leq -\frac{1.6h}{2}(v_{k+1} + v_k)v_k + 4h^2(v_{k+1} + v_k)^2$$
$$= 4h\left\{-\frac{1.6}{8}\left(v_{k+1}v_k + (v_k)^2\right) + h\left[(v_{k+1})^2 + 2v_k v_{k+1} + (v_k)^2\right]\right\}$$
$$= 4h\left[h(v_{k+1})^2 + \left(-\frac{1.6}{8} + 2h\right)v_k v_{k+1} + \left(-\frac{1.6}{8} + h\right)(v_k)^2\right]$$
$$= 4h\left[\left(h - \frac{1.6}{16}\right)(v_{k+1} + v_k)^2 + \frac{1.6}{8}\left(\frac{(v_{k+1})^2 - (v_k)^2}{2}\right)\right].$$

Hence,

$$\left[\frac{1}{2}(v_{k+1})^2 - \frac{1}{2}(v_k)^2\right]\left(1 - \frac{1.6h}{2}\right) - (32\cos x_{k+1} - 32\cos x_k)$$
$$\leq 4h\left(h - \frac{1.6}{16}\right)(v_{k+1} + v_k)^2. \quad (6.19)$$

We now define energy E_k by

$$E_k = \frac{1}{2}(v_k)^2\left(1 - \frac{1.6h}{2}\right) - 32\cos x_k. \quad (6.20)$$

Then, (6.19) can be rewritten as

$$E_{k+1} - E_k \leq 4h \left(h - \frac{1.6}{16} \right) (v_{k+1} + v_k)^2. \tag{6.21}$$

Summing both sides of (6.21) from zero to $n-1$ yields

$$E_n - E_0 \leq 4h \left(h - \frac{1.6}{16} \right) \sum_0^{n-1} (v_{k+1} + v_k)^2.$$

If

$$F_n = -4h \left(h - \frac{1.6}{16} \right) \sum_0^{n-1} (v_{k+1} + v_k)^2, \tag{6.22}$$

then

$$E_n - E_0 \leq -F_n,$$

or

$$E_n + F_n \leq E_0 \quad n \geq 1. \tag{6.23}$$

We now assume that h satisfies two conditions relative to (6.20) and (6.21), namely,

$$1 - \frac{1.6h}{2} > 0 \tag{6.24}$$

and

$$h - \frac{1.6}{16} < 0. \tag{6.25}$$

Conditions (6.24), (6.25) are equivalent to

$$h < \min \left(\frac{1.6}{16}, \frac{2}{1.6} \right) \tag{6.26}$$

We will then show that the desired stability condition is (6.26).
Since (6.24) is valid, it follows from (6.20) that

$$E_k \geq -32 \cos x_k, \tag{6.27}$$

so that

$$-32 \leq E_k. \tag{6.28}$$

Hence, from (6.23),

$$F_n \leq E_0 + 32. \tag{6.29}$$

But, from (6.25), it follows with regard to (6.22) that the sequence $\{F_n\}$ is non-negative, so that

$$0 \leq F_n \leq E_0 + 32. \tag{6.30}$$

Thus, (6.20), (6.23), (6.28) and (6.30) yield

$$-32 \le \frac{1}{2}(v_n)^2 \left(1 - \frac{1.6h}{2}\right) - 32\cos x_n \le E_0 - F_n \le E_0,$$

so that

$$\frac{1}{2}(v_n)^2 \left(1 - \frac{1.6h}{2}\right) \le E_0 + 32.$$

Thus, with the aid of (6.24), it follows that the sequence $\{(v_k)^2\}$ is bounded, and hence v_k is bounded for all k. But the sequence $\{F_n\}$ in (6.22) is, by (6.25) and (6.29) increasing and bounded above, so that it is convergent. Thus, as n goes to infinity,

$$(v_{n+1} + v_n)^2 \to 0,$$

which implies

$$|v_{n+1} + v_n| \to 0. \tag{6.31}$$

Hence, from (6.12), as n goes to infinity,

$$|x_{n+1} - x_n| \to 0. \tag{6.32}$$

But, from (6.16), which is valid for all k, one has

$$v_{k+2} - v_{k+1} = -h(1.6 v_{k+1} + 32 \sin x_{k+1}). \tag{6.33}$$

Summation of (6.16) and (6.33) yields

$$(v_{k+2} + v_{k+1}) - (v_{k+1} + v_k)$$
$$= -h\left[(1.6)(v_{k+1} + v_k) + 32(\sin x_{k+1} + \sin x_k)\right]. \tag{6.34}$$

However, as $k \to \infty$, (6.31) and (6.34) imply

$$(\sin x_{k+1} + \sin x_k) \to 0. \tag{6.35}$$

But the combination (6.32) and (6.35) can only result if x_k converges to an integral multiple of π. Thus, $\{x_k\}$ is bounded and the computation is stable.

Note that the stability condition is independent of initial data.

Example 6.4 For α a fixed, positive constant, consider the initial value problem

$$\ddot{x} + \alpha \dot{x} + 32 \sin x = 0, \quad x_0 = \pi/4, \quad v_0 = 0. \tag{6.36}$$

Note that (6.36) differs from (6.11) only in that the constant 1.6 has been replaced by α. Then, following the same steps given in Example 6.3, but with 1.6 replaced by α, yields the general stability condition

$$h < \min\left(\frac{\alpha}{16}, \frac{2}{\alpha}\right), \tag{6.37}$$

which extends (6.26).

6.2 Instability Analysis

Example 6.5 Consider again initial value problem (6.36). Let us show how to develop a stability condition directly through computational results. Note first that proper lubrication at the joint of the pendulum would imply that α, the damping coefficient in (6.36), is not exceptionally large. Thus, let us assume that $\alpha < 5$. Condition (6.37) then becomes

$$h < \frac{\alpha}{16}. \tag{6.38}$$

Now allow h to have the values $0.025, 0.050, 0.075, 0.100, \ldots, 0.500$. Set $\alpha = 0.8$ and calculate x_k and v_k with

$$x_{k+1} - x_k = \frac{1}{2}h(v_{k+1} + v_k) \tag{6.39}$$

$$v_{k+1} - v_k = -h(\alpha v_k + 32 \sin x_k), \tag{6.40}$$

which are generalizations of (6.15), (6.16). One finds instability for $h \geq 0.050$ and stability for all smaller values of h. Considering, in order, $\alpha = 1.6, 2.4, 3.2,$ and 4.0, one finds instability for h greater than or equal to $0.100, 0.150, 0.200, 0.250$, respectively, and stability for the smaller values of h. Plotting the points $(\alpha, h) = (0.8, 0.50), (1.6, 0.100), (2.4, 0.150), (3.2, 0.200), (4.0, 0.250)$, as shown in Figure 6.1, reveals that α and h are related by the linear equation

$$\alpha = 16h.$$

Thus, the stability condition is $h < \frac{\alpha}{16}$, which is exactly (6.38). Note, also, that the result so derived includes the machine roundoff error.

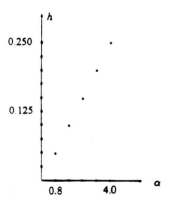

Fig. 6.1 Computational results.

6.3
Numerical Solution of Mildly Nonlinear Autonomous Systems

Invariably, it will be of value to know that a given initial value problem has a unique solution before one attempts to solve it numerically. However, it can also be of value to know the qualitative behavior of the solution, which can serve as a guide to the correctness of a numerical solution. In this section, attention is directed to the latter consideration and the discussion will be in the usual notation of qualitative theory of differential equations.

On $0 < t < \infty$, a system of differential equations of the form

$$\dot{x} = f(x,y) \tag{6.41}$$
$$\dot{y} = g(x,y), \tag{6.42}$$

in which $f(x,y)$ and $g(x,y)$ are independent of t, is called *autonomous*. The points (x_0, y_0) in the plane which satisfy

$$f(x,y) = 0$$
$$g(x,y) = 0.$$

are called *critical* or *equilibrium points* of the system.

The terms critical and equilibrium are meaningful because at such points $\dot{x} = 0$ and $\dot{y} = 0$, so that one's velocity is zero. When one's velocity is zero, then ensuing motion can be in any direction at all, since there is no momentum carrying one forward in any fixed direction. In this sense, one's position is critical.

Example 6.6 The system

$$\dot{x} = 2x - y$$
$$\dot{y} = x - 3y$$

has the single critical point $(0,0)$, which is the unique solution of the system

$$2x - y = 0$$
$$x - 3y = 0.$$

Example 6.7 Consider the system

$$\dot{x} = x(1 - x - y)$$
$$\dot{y} = y(0.75 - y - 0.5x).$$

To find the critical points, we need all the solutions of

$$x(1 - x - y) = 0 \tag{6.43}$$
$$y(0.75 - y - 0.5x) = 0. \tag{6.44}$$

6.3 Numerical Solution of Mildly Nonlinear Autonomous Systems

To find these, we consider all possible solutions of the expanded system

$$x = 0 \tag{6.45}$$
$$1 - x - y = 0 \tag{6.46}$$
$$y = 0 \tag{6.47}$$
$$0.75 - y - 0.5x = 0 \tag{6.48}$$

and discard those which are extraneous to (6.43), (6.44). If $x = 0$, then the possible solutions are $(0,1)$, $(0,0)$, $(0,0.75)$. If $y = 0$, the possible solutions are $(0,0)$, $(1,0)$, $(1.5,0)$. The solution of (6.46) and (6.48) is $(0.5,0.5)$. Inserting each of these points into (6.43), (6.44) yields the solutions $(0,0)$, $(0,0.75)$, (1.0), $(0.5,0.5)$ as the critical points of the given system.

Example 6.8 Consider the system

$$\dot{x} = 67 - x^4 - y^4$$
$$\dot{y} = -35 - x^3 + 3xy.$$

To find the critical points we must solve

$$67 - x^4 - y^4 = 0$$
$$-35 - x^3 + 3xy = 0.$$

Since there is no elementary method for doing this, one can use Newton's method of Section 5.3 to approximate critical points. For example, to four decimal places the result $x = -2.7920$, $y = -1.5803$ was so generated.

Consider system (6.41), (6.42) and let (x_0, y_0) be a critical point.

Definition 6.1 A critical point (x_0, y_0) is called *stable* if for every $\epsilon > 0$ there exists a $\delta > 0$ such that every solution $x = x(t)$, $y = y(t)$, which at $t = 0$ satisfies

$$(x(t) - x_0)^2 + (y(t) - y_0)^2 < \delta$$

exists and satisfies

$$(x(0) - x_0)^2 + (y(0) - y_0)^2 < \epsilon$$

for all $t \geq 0$.

The geometry of a stable critical point in the XY plane is shown in Figure 6.2, where the relationship between x and y is shown graphically. The variables x and y are related through the equations $x = x(t)$, $y = y(t)$, which are parametric equations for the indicated relationship.

It is important to note that Definition 6.1 requires the condition to be valid for *every* solution. This implies that if one produces a general solution with arbitrary constants, then the conditions must be valid for all choices of the constants.

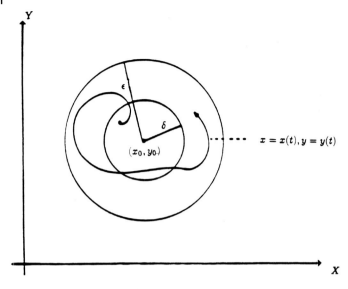

Fig. 6.2 A stable critical point.

Definition 6.2 A critical point (x_0, y_0) is called *asymptotically* stable if it is stable and if, in addition, there exists a $\delta^* > 0$ such that every solution $x = x(t)$, $y = y(t)$ which at $t = 0$ satisfies

$$(x(0) - x_0)^2 + (y(0) - y_0)^2 < \delta^*$$

exists for all $t \geq 0$ and satisfies

$$\lim_{t \to \infty} x(t) = x_0, \quad \lim_{t \to \infty} y(t) = y_0.$$

The geometry of an asymptotically stable critical point is shown in Figure 6.3.

Definition 6.3 Definition 6.3. A critical point which is not stable is called *unstable*.

The geometry of an unstable critical point is shown in Figure 6.4.

Consider now the linear autonomous system

$$\dot{x} = ax + by \tag{6.49}$$
$$\dot{y} = cx + dy, \tag{6.50}$$

in which a, b, c, d are constants which satisfy $ad - bc \neq 0$. This system can be solved exactly. Moreover, $(0,0)$ is the only critical point of the system. Let us then solve (6.49), (6.50) exactly and, from the solution, determine the stability of this critical point. We will do this only because the results will apply later to an important class of nonlinear equations.

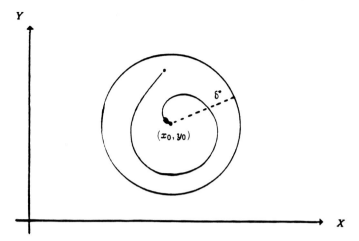

Fig. 6.3 An asymptotically stable critical point.

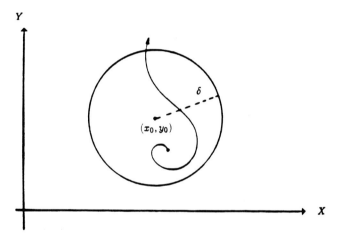

Fig. 6.4 An unstable critical point.

Rewriting (6.49), (6.50) in operator notation yields

$$(D - a)x - by = 0$$
$$-cy + (D - d)y = 0$$

and eliminating the x terms results in

$$\ddot{y} - (a + d)\dot{y} + (ad - bc)y = 0. \tag{6.51}$$

The nature of the solution of (6.51) then depends on the solutions of the algebraic equation

$$m^2 - (a + d)m + (ad - bc) = 0. \tag{6.52}$$

The general solution of (6.51) in terms of the roots m_1, m_2 of (6.52), is given by

$$y = c_1 e^{m_1 t} + c_2 e^{m_2 t}; \quad \text{if } m_1, m_2 \text{ are real and unequal,} \tag{6.53}$$

$$y = c_1 e^{m_1 t} + c_2 t e^{m_1 t}; \quad \text{if } m_1, m_2 \text{ are real and equal,} \tag{6.54}$$

$$y = c_1 e^{At} \cos Bt + c_2 e^{At} \sin Bt; \quad \text{if } m_1 = A + iB, m_2 = A - iB, B \neq 0. \tag{6.55}$$

But, substitution of (6.53)–(6.55) into (6.49) implies, respectively,

$$x = d_1 e^{m_1 t} + d_2 e^{m_2 t} l \tag{6.56}$$

$$x = d_1 e^{m_1 t} + d_2 t e^{m_1 t} \tag{6.57}$$

$$x = d_1 e^{At} \cos Bt + d_2 e^{At} \sin Bt, \tag{6.58}$$

in which the d_1, d_2 are functions of c_1, c_2.

The following theorem now follows directly from (6.53)–(6.58), Definitions 6.1, 6.2, and properties of exponential functions, that is,

$$\lim_{t \to \infty} e^{mt} = \begin{cases} 0, & m < 0 \\ 1, & m = 0 \\ \infty, & m > 0. \end{cases}$$

Theorem 6.1 *Let m_1, m_2 be the roots of (6.52). Then*

(a) *if both roots are real and negative, $(0,0)$ is an asymptotically stable critical point of (6.49),(6.50);*

(b) *if at least one root is real and positive, $(0,0)$ is an unstable critical point;*

(c) *if both roots are complex but have non positive real parts, $(0,0)$ is an stable critical point;*

(d) *if both roots are complex but have negative real parts, then $(0,0)$ is an asymptotically stable critical point;*

(e) *if both roots are complex with positive real part, $(0,0)$ is an unstable critical point.*

Example 6.9 Consider the simple system

$$\dot{x} = -x$$
$$\dot{y} = -y.$$

The critical point of this system is $(0,0)$. Relative to (6.49), (6.50), $a = -1$, $b = 0$, $c = 0$, $d = -1$. Hence, (6.52) reduces to $m^2 + 2m + 1 = 0$, whose roots are $m_1 = m_2 = -1$. Thus, part (a) of Theorem 6.1 implies that $(0,0)$ is an asymptotically stable critical point. And, indeed, since each of the given equations can be integrated directly, the general solution of the system is

$$x(t) = c_1 e^{-t}, y(t) = c_2 e^{-t}. \tag{6.59}$$

From (6.59), it follows that $\lim_{t\to\infty} x(t) = 0$, $\lim_{t\to\infty} y(t) = 0$, independently of c_1, c_2.

Example 6.10 Consider the system

$$\dot{x} = -3x + 4y \tag{6.60}$$
$$\dot{y} = -2x + 3y. \tag{6.61}$$

The critical point is $(0,0)$. Since $a = -3$, $b = 4$, $c = -2$, $d = 3$, then (6.52) reduces to $m^2 - 1 = 0$. Hence $m_1 = 1$, $m_2 = -1$, so that part (b) of Theorem 6.1 implies that $(0,0)$ is an unstable critical point. Indeed, the general solution of the system is

$$x(t) = c_1 e^t + 2c_2 e^{-t}$$
$$y(t) = c_1 e^t + c_2 e^{-t}.$$

For c_1 not equal to zero, then, $\lim_{t\to\infty} e^t$ is unbounded, which implies instability.

As can be seen from the discussion above, there is no real value in determining whether a critical point is or is not stable when system (6.49), (6.50) is linear with constant coefficients, since in all such cases the exact solution can be found readily. The value of qualitative theory lies in its being able to provide the means to determine if a critical point is stable or not when system (6.41), (6.42) is nonlinear.

Stability analysis for nonlinear equations can be exceptionally difficult, and, indeed, not always possible. However, it is relatively easy to apply to a class of equations which are called *mildly* nonlinear. These are defined as follows. Consider the system

$$\dot{x} = ax + by + F(x, y) \tag{6.62}$$
$$\dot{y} = cx + dy + G(x, y), \tag{6.63}$$

in which a, b, c, d are constants that satisfy $ad - bc \neq 0$ and $F(0,0) = 0$, $G(0,0) = 0$. It follows that $(0,0)$ is a critical point of the system. Assume further that $F(x, y)$, $G(x, y)$ are continuous, have continuous first partials, and satisfy

$$\lim_{x^2+y^2 \to 0} \frac{F(x,y)}{(x^2+y^2)^{\frac{1}{2}}} = 0, \qquad \lim_{x^2+y^2 \to 0} \frac{G(x,y)}{(x^2+y^2)^{\frac{1}{2}}} = 0. \tag{6.64}$$

Then the system (6.62), (6.63) is called mildly nonlinear.

Example 6.11 The system

$$\dot{x} = -x + y + (x^2 + y^2)$$
$$\dot{y} = -2x - (x^2 + y^2)^{3/2}$$

has the point $(0,0)$ for a critical point. Further it is mildly nonlinear because $a = -1$, $b = 1$, $c = 0$, $d = -2$ and $ad - bc = 2$, while

$$\lim_{x^2+y^2 \to 0} \frac{x^2 + y^2}{(x^2 + y^2)^{\frac{1}{2}}} = 0, \qquad \lim_{x^2+y^2 \to 0} \frac{-(x^2 + y^2)^{3/2}}{(x^2 + y^2)^{\frac{1}{2}}} = 0.$$

Example 6.12 The system

$$\dot{x} = -3x + 4y + (x^2 - y^2)$$
$$\dot{y} = -2x + 3y - xy$$

has $(0,0)$ for a critical point. Further, it is mildly nonlinear because $a = -3$, $b = 4$, $c = -2$, $d = 3$, so that $ad - bc = -1$, and, using polar coordinates

$$\lim_{x^2+y^2 \to 0} \frac{x^2 - y^2}{(x^2 + y^2)^{\frac{1}{2}}} = \lim_{r^2 \to 0} \frac{r^2(\cos^2\theta - \sin^2\theta)}{r} = 0$$

$$\lim_{x^2+y^2 \to 0} \frac{-xy}{(x^2 + y^2)^{\frac{1}{2}}} = \lim_{r^2 \to 0} \frac{-r^2 \cos\theta \sin\theta}{r} = 0,$$

independently of θ. Note, of course, that $r^2 \to 0$ implies $r \to 0$.

For mildly nonlinear autonomous systems, the following theorem is available (Cesari (1959)). Its significance lies in the fact that it allows one to determine the stability of a nonlinear system through the stability of a related linear system.

Theorem 6.2 *Consider the mildly nonlinear system (6.62), (6.63). Then,*

(a) *the critical point $(0,0)$ is asymptotically stable if $(0,0)$ is asymptotically stable for linear system (6.49), (6.50);*

(b) *the critical point $(0,0)$ is unstable if $(0,0)$ is unstable for the linear system (6.49), (6.50).*

Example 6.13 Consider the mildly nonlinear system

$$\dot{x} = -x + y + (x^2 + y^2) \qquad (6.65)$$
$$\dot{y} = -2x - (x^2 + y^2)^{3/2} \qquad (6.66)$$

Since $a = -1$, $b = 1$, $c = 0$, $d = -2$, equation (6.52) reduces to $m^2 + 3m + 2 = 0$, whose roots are $m_1 = -2$, $m_2 = -1$. Thus, by Theorem 6.1, the critical point $(0,0)$ is asymptotically stable for linear system (6.49), (6.50). Then, by Theorem 6.2, the critical point $(0,0)$ of (6.65), (6.66) is asymptotically stable.

6.3 Numerical Solution of Mildly Nonlinear Autonomous Systems

The result of the above example is very useful if one now wishes to approximate a solution of (6.65), (6.66) from given initial conditions. For this purpose, let us consider the initial data

$$x(0) = 0.1, \quad y(0) = 0.1, \tag{6.67}$$

and solve (6.65), (6.66) using a third-order Taylor expansion.

The program was run for 100 000 steps with $h = 0.0001$. The resulting x and y values are recorded every 10000 steps in Table 6.1. The derivative values, every 10 000 steps, are recorded in Table 6.2. The convergence of the points (x, y) to the critical point $(0, 0)$ is in agreement with Definition 6.2 of a critical point which is asymptotically stable.

Table 6.1 Solution of the system (6.65), (6.66).

x	y
$0.655247576E - 01$	$0.132026698E - 01$
$0.283525486E - 01$	$0.175489037E - 02$
$0.110349094E - 01$	$0.235367818E - 03$
$0.414305763E - 02$	$0.317349882E - 04$
$0.153554187E - 02$	$0.428871104E - 05$
$0.566441150E - 03$	$0.580102909E - 06$
$0.208591628E - 03$	$0.784928200E - 07$
$0.767649458E - 04$	$0.106220711E - 07$
$0.282440859E - 04$	$0.143750230E - 08$
$0.103909383E - 04$	$0.194542853E - 09$

Table 6.2 Derivative values for the solution of the system (6.65), (6.66).

$\dot{x}$	$\ddot{x}$	$\dddot{x}$	$\dot{y}$	$\ddot{y}$	$\dddot{y}$
$-0.48E - 01$	$0.14E - 01$	$0.49E - 01$	$-0.27E - 01$	$0.54E - 01$	$-0.11E + 00$
$-0.26E - 01$	$0.21E - 01$	$-0.11E - 01$	$-0.35E - 02$	$0.71E - 02$	$-0.14E - 01$
$-0.11E - 01$	$0.10E - 01$	$-0.86E - 02$	$-0.47E - 03$	$0.95E - 03$	$-0.19E - 02$
$-0.41E - 02$	$0.40E - 02$	$-0.38E - 02$	$-0.64E - 04$	$0.13E - 03$	$-0.26E - 03$
$-0.15E - 02$	$0.15E - 02$	$-0.15E - 02$	$-0.86E - 05$	$0.17E - 04$	$-0.34E - 04$
$-0.57E - 03$	$0.56E - 03$	$-0.56E - 03$	$-0.12E - 05$	$0.23E - 05$	$-0.46E - 05$
$-0.21E - 03$	$0.21E - 03$	$-0.21E - 03$	$-0.16E - 06$	$0.31E - 06$	$-0.63E - 06$
$-0.77E - 04$	$0.77E - 04$	$-0.77E - 04$	$-0.21E - 07$	$0.42E - 07$	$-0.85E - 07$
$-0.28E - 04$	$0.28E - 04$	$-0.28E - 04$	$-0.29E - 08$	$0.58E - 08$	$-0.12E - 07$
$-0.10E - 04$	$0.10E - 04$	$-0.10E - 04$	$-0.39E - 09$	$0.78E - 09$	$-0.16E - 08$

The computer program used is given generically as follows.

Algorithm 8 Program Stable

Step 1. Set a counter $K = 1$.
Step 2. Set a time step h.
Step 3. Set an initial time t.
Step 4. Set initial values x and y.
Step 5. Define $x_1 = \dot{x}$, $x_2 = \ddot{x}$, $x_3 = \dddot{x}$, $y_1 = \dot{y}$, $y_2 = \ddot{y}$, $y_3 = \dddot{y}$,
Step 6. Calculate
$$d = (x^2 + y^2)^{\frac{1}{2}}$$
$$x_1 = -x + y + d^2$$
$$y_1 = -2y - d^3$$
$$x_2 = -x_1 + y_1 + 2xx_1 + 2yy_1$$
$$y_2 = -2y_1 - 3d(xx_1 + yy_1)$$
$$x_3 = -x_2 + y_2 + 2(x_1)^2 + 2xx_2 + 2(y_1)^2 + 2yy_2$$
$$y_3 = -2y_2 - 1.5(xx_1 + yy_1)^2(d)^{-1} - 3d(x_1^2 + xx_2 + y_1^2 + yy_2).$$
Step 7. Calculate x at $t + h$ and y at $t + h$ by
$$x(t + h) = x + hx_1 + \tfrac{1}{2}h^2 x_2 + \tfrac{1}{6}h^3 x_3$$
$$y(t + h) = y + hy_1 + \tfrac{1}{2}h^2 y_2 + \tfrac{1}{6}h^3 y_3.$$
Step 8. Increase the counter from K to $K + 1$.
Step 9. Set $x = x(x + h)$, $y = y(t + h)$.
Step 10. Repeat Steps 6 – 9.
Step 11. Continue until $K = 100\,000$.
Step 12. Stop the calculation.

6.4 Exercises

6.1 For what values of h is Euler's method stable for each of the following initial value problems?

(a) $y' = -y$, $y(0) = 1$
(b) $y' = -10y$, $y(0) = 2$
(c) $y' = -100y$, $y(0) = 0$
(d) $y' = -1000y$, $y(0) = -1$
(e) $y' = y$, $y(0) = 1$
(f) $y' = 100y$, $y(0) = 0$
(g) $y' = 10000y$, $y(0) = -1$.

6.2 Which of the following are stable difference equations and which are not?

(a) $y_{i+2} - 3y_{i+1} + 2y_i = 0$
(b) $2y_{i+2} - 3y_{i+1} + y_i = 0$
(c) $2y_{i+2} + 3y_{i+1} + y_i = 0$
(d) $4y_{i+2} + 4y_{i+1} + 4y_i = 0$
(e) $y_{i+2} + 4y_{i+1} + y_i = 0$
(f) $4y_{i+2} + y_{i+1} + 4y_i = 0$
(g) $4y_{i+2} - 3y_{i+1} + y_i = 0$
(h) $y_{i+2} - 3y_{i+1} + 4y_i = 0$
(i) $4y_{i+2} - 5y_{i+1} + y_i = 0$
(j) $4y_{i+2} + y_{i+1} - 5y_i = 0$
(k) $y_{i+2} + 10y_{i+1} - 11y_i = 0$
(l) $11y_{i+2} - 10y_{i+1} - y_i = 0$
(m) $y_{i+2} - y_i = 0$
(n) $y_{i+2} + y_{i+1} = 0$.

6.3 Using the recursion formulas (6.12)–(6.14), determine a general stability condition for the initial value problem

$$\ddot{x} + \alpha\dot{x} + \beta \sin x = 0, \quad x(0) = \gamma, \quad \dot{x}(0) = 0, \quad \alpha > 0, \; \beta > 0, \; \gamma > 0.$$

6.4 Show that the use of the energy conserving methodology of Chapter 5 is always stable for the initial value problem in Exercise 3 if $\alpha = 0$.

6.5 Show that the energy conserving methodology of Chapter 5 is always stable for coulombic repulsion provided the energy is negative.

7
Numerical Solution of Tridiagonal Linear Algebraic Systems and Related Nonlinear Systems

7.1
Introduction

The numerical method to be developed for the solution of linear boundary value problems requires the ability to solve systems of tridiagonal, linear algebraic equations. For the nonlinear boundary problems to be considered, the resulting equations will be nonlinear, but will be related structurally to tridiagonal systems. In this summary chapter, we describe basic theorems and popular methods for solving such systems and only provide references for the related proofs.

7.2
Tridiagonal Systems

For $n \geq 2$, the general linear algebraic system of n equations in the n unknowns $x_1, x_2, x_3, \ldots, x_n$ is

$$\begin{aligned}
a_{11}x_1 + a_{12}x_2 + a_{13}x_3 + \ldots + a_{1n}x_n &= b_1 \\
a_{21}x_1 + a_{22}x_2 + a_{23}x_3 + \ldots + a_{2n}x_n &= b_2 \\
a_{31}x_1 + a_{32}x_2 + a_{33}x_3 + \ldots + a_{3n}x_n &= b_3 \\
\vdots \quad \vdots \quad \vdots \quad \vdots \quad \vdots \quad &\vdots \\
a_{n1}x_1 + a_{n2}x_2 + a_{n3}x_3 + \ldots + a_{nn}x_n &= b_n.
\end{aligned} \qquad (7.1)$$

Numerical Solution of Ordinary Differential Equations for Classical, Relativistic and Nano Systems. Donald Greenspan
Copyright © 2006 WILEY-VCH Verlag GmbH & Co. KGaA, Weinheim
ISBN: 3-527-40610-7

If matrix A and vectors x and b are defined by

$$A = \begin{bmatrix} a_{11} & a_{12} & a_{13} & \cdots & a_{1n} \\ a_{21} & a_{22} & a_{23} & \cdots & a_{2n} \\ a_{31} & a_{32} & a_{33} & \cdots & a_{3n} \\ \vdots & \vdots & \vdots & \vdots & \vdots \\ a_{n1} & a_{n2} & a_{n3} & \cdots & a_{nn} \end{bmatrix}, \quad x = \begin{bmatrix} x_1 \\ x_2 \\ x_3 \\ \vdots \\ x_n \end{bmatrix}, \quad b = \begin{bmatrix} b_1 \\ b_2 \\ b_3 \\ \vdots \\ b_n \end{bmatrix}, \qquad (7.2)$$

then system (7.1) can be written compactly as (see the Appendix):

$$Ax = b. \qquad (7.3)$$

The system (7.1), or, equivalently, (7.3), is said to be tridiagonal if all the elements are zero except a_{ii}, $a_{j,j+1}$, $a_{j+1,j}$, for $i = 1, 2, 3, \ldots, n;\ j = 1, 2, 3, \ldots, n-1$, and none of these is zero.

An example of a tridiagonal system with $n = 5$ is

$$\begin{aligned}
-4x_1 + x_2 &= 1 \\
x_1 - 3x_2 + x_3 &= 7 \\
x_2 - 5x_3 + x_4 &= 5 \\
x_3 - 6x_4 + x_5 &= 0 \\
x_4 - 7x_5 &= 1.
\end{aligned}$$

The term tridiagonal is most appropriate because matrix A has the particular form

$$A = \begin{bmatrix} a_{11} & a_{12} & & & & & \\ a_{21} & a_{22} & a_{23} & & & 0 & \\ & a_{32} & a_{33} & a_{34} & & & \\ & & \cdots & \cdots & \cdots & & \\ & & & \cdots & \cdots & \cdots & \\ & & & & \cdots & \cdots & \cdots \\ & 0 & & a_{n-1,n-2} & a_{n-1,n-1} & a_{n-1,n} \\ & & & & a_{n,n-1} & a_{n,n} \end{bmatrix}, \qquad (7.4)$$

in which all the elements are zero except those on the main diagonal, the super diagonal (just above the main diagonal) and the sub diagonal (just below the main diagonal).

Of practical importance in ensuring that a tridiagonal system has a unique solution is the following sufficiency theorem (Greenspan and Casulli (1988)).

Theorem 7.1 *Let system (7.3) be tridiagonal. Let the main diagonal elements be negative, while the sub diagonal and super diagonal elements are positive, that is,*

$$a_{ii} < 0, \quad i = 1, 2, 3, \ldots, n \qquad (7.5)$$
$$a_{i,i+1} > 0, \quad i = 1, 2, 3, \ldots, n-1 \qquad (7.6)$$
$$a_{i+1,i} > 0, \quad i = 1, 2, 3, \ldots, n-1. \qquad (7.7)$$

Further, let the main diagonal elements dominate the matrix, that is, let the absolute value of the main diagonal elements be greater than or equal to the sum of all other row elements, with strict inequality for at least one row. Precisely, then, let

$$-a_{11} \geq a_{12} \tag{7.8}$$

$$-a_{nn} \geq a_{n,n-1} \tag{7.9}$$

$$-a_{ii} \geq a_{i,i-1} + a_{i,i+1}, \quad i = 2, 3, \ldots, n-1, \tag{7.10}$$

with strict inequality holding for at least one of (7.8)–(7.10). Then the resulting linear algebraic system has one and only one solution.

Example 7.1 The system

$$
\begin{aligned}
-2x_1 + x_2 &= 7 \\
x_1 - 2x_2 + x_3 &= 1 \\
x_2 - 2x_3 + x_4 &= 0 \\
x_3 - 2x_4 + x_5 &= 0 \\
x_4 - 2x_5 &= 2
\end{aligned}
$$

has one and only one solution.

Example 7.2 The system

$$
\begin{aligned}
-x_1 + x_2 &= 1 \\
x_1 - 2x_2 + x_3 &= 1 \\
x_2 - 2x_3 + x_4 &= 1 \\
x_3 - 3x_4 + x_5 &= 7 \\
x_4 - x_5 &= 6
\end{aligned}
$$

has a unique solution.

Example 7.3 Theorem 7.1 provides no information about the number of solutions of the system

$$
\begin{aligned}
-x_1 + x_2 &= 1 \\
x_1 - 2x_2 + x_3 &= 1 \\
x_2 - 2x_3 + x_4 &= 1 \\
x_3 - 2x_4 + x_5 &= 0 \\
x_4 - x_5 &= 1.
\end{aligned}
$$

The availability of a theorem like Theorem 7.1 before one begins computations cannot be underestimated. If a system were to have no solution, computations can yield nonsense numbers. If a system were to have more than one solution, certain computational techniques could drift from one solution to another.

Note that Theorem 7.1 will serve to guide one's intuition later in developing viable numerical methodology for solving boundary value problems.

7.3
The Direct Method

In this section we give, by means of a generic algorithm, a popular method, called the direct method, for solving tridiagonal systems which satisfy the conditions of Theorem 7.1 (Greenspan and Casulli (1988)). The algorithm proceeds as follows:

Algorithm 9 Direct Method

Step 1. Store the constants
$$a_{ii}, \quad i = 1, 2, 3, \ldots, n$$
$$a_{i,i+1}, \quad i = 1, 2, 3, \ldots, n-1$$
$$a_{i+1,i}, \quad i = 1, 2, 3, \ldots, n-1.$$
$$b_i, \quad i = 1, 2, 3, \ldots, n$$

Step 2. Determine $p_i, i = 1, 2, 3, \ldots, n$; and $q_j, j = 2, 3, \ldots, n$ by
$$p_1 = a_{11}$$
$$q_{j-1} = a_{j-1}/p_{j-1}; \quad j = 2, 3, \ldots, n$$
$$p_j = a_{jj} - a_{j,j-1} q_{j-1}; \quad j = 2, 3, \ldots, n$$

Step 3. Determine the constants $z_i, \quad i = 1, 2, 3, \ldots, n$, from
$$z_1 = b_1/p_1$$
$$z_j = (b_j - a_{j,j-1} z_{j-1})/p_j; \quad j = 2, 3, \ldots, n.$$

Step 4. Determine the solution vector x by
$$x_n = z_n$$
$$x_j = z_j - q_j x_{j+1}, \quad j = n-1, n-2, \ldots, 3, 2, 1.$$

Example 7.4 Consider the system

$$\begin{aligned}
-2x_1 + x_2 &= 1 \\
x_1 - 2x_2 + x_3 &= 0 \\
x_2 - 2x_3 + x_4 &= 0 \\
x_3 - 2x_4 + x_5 &= 0 \\
x_4 - 2x_5 &= 0.
\end{aligned}$$

Then, $a_{ii} = -2, i = 1, 2, 3, 4, 5; a_{j,j+1} = a_{j+1,j} = 1, j = 1, 2, 3, 4; b_1 = 1, b_2 = b_3 = b_4 = b_5 = 0.$ From Step 2,

$$p_1 = -2, \qquad q_1 = 1/p_1 = -\frac{1}{2}$$
$$p_2 = -2 - \left(-\frac{1}{2}\right) = -\frac{3}{2}, \quad q_2 = 1/p_2 = -\frac{2}{3}$$
$$p_3 = -\frac{4}{3}, \qquad q_3 = -\frac{3}{4}$$
$$p_4 = -\frac{5}{4}, \qquad q_4 = -\frac{4}{5}$$
$$p_5 = -\frac{6}{5}.$$

From Step 3,

$$z_1 = b_1/p_1 = -\frac{1}{2}$$

$$z_2 = \frac{0 - 1(z_1)}{p_2} = -\frac{1}{3}$$

$$z_3 = -\frac{1}{4}$$

$$z_4 = -\frac{1}{5}$$

$$z_5 = -\frac{1}{6}.$$

From Step 4,

$$x_5 = z_5 = -\frac{1}{6}$$

$$x_4 = z_4 - x_5\left(-\frac{4}{5}\right) = -\frac{1}{5} - \left(-\frac{1}{6}\right)\left(-\frac{4}{5}\right) = -\frac{1}{3}$$

$$x_3 = -\frac{1}{2}$$

$$x_2 = -\frac{2}{3}$$

$$x_1 = -\frac{5}{6}.$$

Thus, the solution is

$$x_1 = -\frac{5}{6}, \quad x_2 = -\frac{2}{3}, \quad x_3 = -\frac{1}{2}, \quad x_4 = -\frac{1}{3}, \quad x_5 = -\frac{1}{6},$$

and this solution is verified to be correct by substitution into the given system.

In practice, when using a 32-bit machine, the direct method seems to work well for systems of up to 1500 equations, after which roundoff error tends to accumulate.

7.4
The Newton–Lieberstein Method

The direct method of Section 7.3 is called direct because the algorithm terminates in a fixed number of steps and this number of steps can be determined *a priori*. In this section we will develop a different type of method, called an iterative method, for which the number of steps cannot be determined *a priori*. Unlike the direct method, however, it will be applicable to nonlinear systems and to systems with more than 1500 equations. It is a variation of Newton's method called the Newton–Lieberstien method (Lieberstein (1959)), and proceeds as follows.

Consider the system

$$
\begin{aligned}
a_{11}x_1 + a_{12}x_2 &= f_1(x_1) \\
a_{21}x_1 + a_{22}x_2 + a_{23}x_3 &= f_2(x_2) \\
a_{32}x_2 + a_{33}x_3 + a_{34}x_4 &= f_3(x_3) \\
&\ \vdots \\
a_{n,n-1}x_{n-1} + a_{nn}x_n &= f_n(x_n)
\end{aligned}
$$

in which the matrix of coefficients is (7.4). Let this matrix satisfy the conditions of Theorem 7.1. If $f_1(x_1) = b_1$, $f_2(x_2) = b_2$, ..., $f_n(x_n) = b_n$, the system under consideration is the same as the system in Theorem 7.1. If the $f_i(x_i)$ are not all constants, and if $f'_i(x_i) \geq 0$, then the system is called mildly nonlinear. All such mildly nonlinear systems have unique solutions (Ortega and Rheinboldt (1981)).

Example 7.5

$$
\begin{aligned}
-2x_1 + x_2 &= e^{x_1} \\
x_1 - 2x_2 + x_3 &= e^{x_2} \\
x_2 - 2x_3 + x_4 &= e^{x_3} \\
x_3 - 2x_4 + x_5 &= e^{x_4} \\
x_4 - 2x_5 &= e^{x_5}
\end{aligned}
$$

is a mildly nonlinear system which has a unique solution.

The Newton Lieberstein method for solving a mildly nonlinear system is given in the following generic algorithm.

Algorithm 10 Newton–Lieberstein

Step 1. Guess an initial vector $x_1^{(0)}, x_2^{(0)}, x_3^{(0)}, \ldots, x_n^{(0)}$ and a value ω in the range $0 < \omega < 2$.

Step 2. For $k = 0, 1, 2, 3, \ldots$, iterate with the formulas

$$x_1^{(k+1)} = x_1^{(k)} - \omega \frac{a_{11}x_1^{(k)} + a_{12}x_2^{(k)} - f_1(x_1^{(k)})}{a_{11} - f'_1(x_1^{(k)})}$$

$$x_2^{(k+1)} = x_2^{(k)} - \omega \frac{a_{21}x_1^{(k+1)} + a_{22}x_2^{(k)} + a_{23}x_3^{(k)} - f_2(x_2^{(k)})}{a_{22} - f'_2(x_2^{(k)})}$$

$$x_3^{(k+1)} = x_3^{(k)} - \omega \frac{a_{32}x_2^{(k+1)} + a_{33}x_3^{(k)} + a_{34}x_4^{(k)} - f_3(x_3^{(k)})}{a_{33} - f'_3(x_3^{(k)})}$$

$$\vdots$$

$$x_{n-1}^{(k+1)} = x_{n-1}^{(k)} - \omega \frac{a_{n-1,n-2}x_{n-2}^{(k+1)} + a_{n-1,n-1}x_{n-1}^{(k)} + a_{n-1,n}x_n^{(k)} - f_{n-1}(x_{n-1}^{(k)})}{a_{n-1,n-1} - f'_{n-1}(x_{n-1}^{(k)})}$$

$$x_n^{(k+1)} = x_n^{(k)} - \omega \frac{a_{n,n-1}x_{n-1}^{(k+1)} + a_{nn}x_n^{(k)} - f_n(x_n^{(k)})}{a_{nn} - f'_n(x_n^{(k)})}$$

Step 3. Continue the iteration until, for prescribed convergence tolerance ϵ, one has

$$\left|x_i^{(k+1)} - x_i^{(k)}\right| < \epsilon, \quad i = 1, 2, 3, \ldots, n.$$

Step 4. Substitute the values $x_1^{(k+1)}, x_2^{(k+1)}, x_3^{(k+1)}, \ldots, x_n^{(k+1)}$ into the original system of equations to check that they are an approximate solution.

Example 7.6 Consider the mildly nonlinear system

$$\begin{aligned}
-50x_1 + 25x_2 &= e^{x_1} \\
25x_1 - 50x_2 + 25x_3 &= e^{x_2} \\
25x_2 - 50x_3 + 25x_4 &= e^{x_3} \\
25x_3 - 50x_4 &= e^{x_4}.
\end{aligned}$$

For this system the iteration formulas are

$$x_1^{(k+1)} = x_1^{(k)} - \omega \frac{-50x_1^{(k)} + 25x_2^{(k)} - e^{x_1^{(k)}}}{-50 - e^{x_1^{(k)}}}$$

$$x_2^{(k+1)} = x_2^{(k)} - \omega \frac{25x_1^{(k+1)} - 50x_2^{(k)} + 25x_3^{(k)} - e^{x_2^{(k)}}}{-50 - e^{x_2^{(k)}}}$$

$$x_3^{(k+1)} = x_3^{(k)} - \omega \frac{25x_2^{(k+1)} - 50x_3^{(k)} + 25x_4^{(k)} - e^{x_3^{(k)}}}{-50 - e^{x_3^{(k)}}}$$

$$x_4^{(k+1)} = x_4^{(k)} - \omega \frac{25x_3^{(k+1)} - 50x_4^{(k)} - e^{x_4^{(k)}}}{-50 - e^{x_4^{(k)}}}$$

With zero initial guess and $\omega = 1.3$, the formulas converged to a numerical solution in only eight iterations, which, to four decimal places, is

$$x_1 = -0.0731, \; x_2 = -0.1089, \; x_3 = -0.1089, \; x_4 = -0.0731.$$

In practice, for very large systems of equations, one usually tries $\omega = 1.0$, 1.3, 1.5, 1.7 for ten iterations each, and then chooses that value of ω which is yielding the most rapid convergence for the remainder of the iterations.

When applied to linear systems, the Newton–Lieberstein method is the same as the successive overrelaxtion method (SOR).

7.5
Exercises

7.1 Find the numerical solution of each of the following tridiagonal systems and check each answer:

(a) $-2x_1 + x_2 = 5$
$x_1 - 2x_2 + x_3 = -4$
$x_{j-1} - 2x_j + x_{j+1} = 0, \quad j = 3, 4, 5, \ldots, 97, 98$
$x_{98} - 2x_{99} + x_{100} = -8$
$x_{99} - 2x_{100} = 13$

(b) $-2x_1 + x_2 = 5$
$x_1 - 3x_2 + x_3 = -4$
$x_{j-1} - 4x_j + x_{j+1} = 0, \quad j = 3, 4, 5, \ldots, 197, 198$
$x_{198} - 3x_{199} + x_{200} = -8$
$x_{199} - 2x_{200} = 13$

(c) $3x_1 - x_2 = 1$
$-x_{j-1} + 3x_j - x_{j+1} = 0, \quad j = 2, 3, 4, 5, \ldots, 197, 198, 199$
$-x_{199} + 3x_{200} = 1$

(d) $-2x_1 + x_2 = 0.5$
$x_{j-1} - 2x_j + x_{j+1} = \dfrac{j}{(j+1)}, \quad j = 2, 3, 4, 5, \ldots, 997, 998, 999$
$x_{999} - 2x_{1000} = 1.$

7.2 Find the numerical solution of each of the following systems and check each answer:

(a) $-2x_1 + x_2 = 5e^{x_1}$
$x_1 - 2x_2 + x_3 = 4e^{x_2}$
$x_{j-1} - 2x_j + x_{j+1} = e^{x_j}, \quad j = 3, 4, 5, \ldots, 97, 98$
$x_{98} - 2x_{99} + x_{100} = 8e^{x_{99}}$
$x_{99} - 2x_{100} = 13e^{x_{100}}$

(b) $-2x_1 + x_2 = 5(x_1)^3$
$x_1 - 3x_2 + x_3 = 4(x_2)^3$
$x_{j-1} - 4x_j + x_{j+1} = (x_j)^3, \quad j = 3, 4, 5, \ldots, 197, 198$
$x_{198} - 3x_{199} + x_{200} = 8(x_{199})^3$
$x_{199} - 2x_{200} = 13(x_{200})^3$

7.3 Prove Theorem 7.1 for $n = 5$.

7.4 Prove Theorem 7.1.

8
Approximate Solution of Boundary Value Problems

8.1
Introduction

Physically, a boundary value problem for a second-order differential equation is a problem in which one knows one's position at time $x = a$, knows where one wants to be at a later time $x = b$, and has one's motion prescribed by a given differential equation in the time interval $a < x < b$. Mathematically, a boundary value problem is defined by the differential equation

$$y'' = f(x, y, y'), \quad a < x < b, \quad (8.1)$$

and the boundary conditions

$$y(a) = \alpha, \; y(b) = \beta, \quad a < b \quad (8.2)$$

in which a, b, α, β are constants. One must find a solution $y(x)$ of (8.1) which is continuous on $a \leq x \leq b$ and satisfies the boundary conditions (8.2).

8.2
Approximate Differentiation

Thus far we have introduced three approximations for derivatives at a grid point x_i, namely,

$$y'_i = \frac{y_{i+1} - y_i}{h} \quad \text{(two point forward)} \quad (8.3)$$

$$y'_i = \frac{y_i - y_{i-1}}{h} \quad \text{(two point backward)} \quad (8.4)$$

$$y'_i = \frac{y_{i+1} - y_{i-1}}{2h} \quad \text{(two point central)}. \quad (8.5)$$

8 Approximate Solution of Boundary Value Problems

We will require one additional formula, that is, a formula for $y''(x_k) = y_k''$. For this we use simply (8.3) and (8.4) and have the approximation

$$y_i'' = \frac{\frac{y_{i+1}-y_i}{h} - \frac{y_i-y_{i-1}}{h}}{h},$$

or,

$$y_i'' = \frac{y_{i-1} - 2y_i + y_{i+1}}{h^2} \qquad \text{(three point central).} \qquad (8.6)$$

Though we have derived difference approximations (8.3)–(8.6) intuitively, each can be derived rigorously using finite Taylor expansions (Greenspan and Casulli (1988)).

8.3
Numerical Solution of Boundary Value Problems Using Difference Equations

One of the most efficient ways to solve boundary value problems numerically is to approximate the derivatives of the differential equation by differences and then to solve a system of algebraic or transcendental equations using the methods of Chapter 7. Let us illustrate this methodology first with a simple example and then proceed to more complex problems.

Consider the boundary value problem

$$y'' = 0 \qquad (8.7)$$
$$y(0) = -1, \quad y(6) = 5. \qquad (8.8)$$

The exact solution is, of course,

$$y = x - 1. \qquad (8.9)$$

Proceeding numerically for illustrative purposes only, let us divide $0 \leq x \leq 6$ into n equal parts. For example, let $n = 6$. Thus, $h = 1$, $x_0 = 0$, $x_1 = 1$, $x_2 = 2$, $x_3 = 3$, $x_4 = 4$, $x_5 = 5$, $x_6 = 6$. On these grid points let $y_i = y(x_i)$. From (8.8), then, $y_0 = -1$, $y_6 = 5$. The problem is to determine y_1, y_2, y_3, y_4, y_5. This is accomplished by first approximating the differential equation at each of the interior grid points x_1, x_2, x_3, x_4, x_5 as follows. Let x_i be an arbitrary interior grid point. At x_i substitute the difference approximation (8.6) for y'' in (8.7) to yield the approximating difference equation

$$\frac{y_{i-1} - 2y_i + y_{i+1}}{h^2} = 0$$

or, since $h = 1$,

$$y_{i-1} - 2y_i + y_{i+1} = 0. \qquad (8.10)$$

8.3 Numerical Solution of Boundary Value Problems Using Difference Equations

Writing (8.10) at each interior grid point, in order, for $i = 1, 2, 3, 4, 5$, yields

$$y_0 - 2y_1 + y_2 = 0$$
$$y_1 - 2y_2 + y_3 = 0$$
$$y_2 - 2y_3 + y_4 = 0$$
$$y_3 - 2y_4 + y_5 = 0$$
$$y_4 - 2y_5 + y_6 = 0.$$

Inserting the boundary values $y_0 = -1$, $y_6 = 5$ into this system then implies

$$
\begin{aligned}
-2y_1 + y_2 &= 1 \\
y_1 - 2y_2 + y_3 &= 0 \\
y_2 - 2y_3 + y_4 &= 0 \\
y_3 - 2y_4 + y_5 &= 0 \\
y_4 - 2y_5 &= -5
\end{aligned}
$$

which is a tridiagonal system which satisfies all the conditions of Theorem 7.1. Thus, the solution exists and is unique and can be found by either the direct method or the Newton–Lieberstein method of Chapter 7. The solution is

$$y_1 = 0, \quad y_2 = 1, \quad y_3 = 2, \quad y_4 = 3, \quad y_5 = 4, \qquad (8.11)$$

which, in addition to $y_0 = -1$, $y_6 = 5$ is called the numerical solution. Note, incidentally, that the numerical solution (8.11) and the exact solution (8.9) are identical at the grid points. This is rarely the case and is valid in this case only because of the simplicity of the problem (8.7), (8.8).

The basic ideas of the above example will be extended now to the general linear boundary value problem as follows.

Let I be the open interval $a < x < b$ and let α, β be constants. Let $P(x), Q(x), R(x)$ be continuous on $a \leq x \leq b$ and consider the boundary value problem

$$y'' + P(x)y' + Q(x)y = R(x), \quad x \in I \qquad (8.12)$$
$$y(a) = \alpha, \quad y(b) = \beta. \qquad (8.13)$$

For numerical purposes, it is, of course, most desirable to know, *a priori*, that problem (8.12), (8.13) has a unique solution. For this purpose, we assume, in addition, that

$$Q(x) \leq 0, \quad a \leq x \leq b, \qquad (8.14)$$

which is sufficient to ensure existence and uniqueness of the solution (Courant and Hilbert (1962)).

Except special cases, numerical methodology is required for (8.12), (8.13). The following general algorithm, which will be discussed after an illustrative example is given, will be used.

Algorithm 11 Linear Boundary Value Problem Solution

Step 1. Determine a nonnegative constant M which satisfies $|P(x)| \leq M$ on $a \leq x \leq b$. Divide $a \leq x \leq b$ into n equal parts by the grid points $a = x_0, x_1, x_2, \ldots, x_n = b$, but let the grid size h satisfy the condition

$$Mh < 2. \quad (8.15)$$

Step 2. Let $y_i = y(x_i)$, $i = 0, 1, 2, 3, \ldots, n$ and note that $y_0 = \alpha$, $y_n = \beta$.

Step 3. At each interior grid point x_i, $i = 1, 2, 3, \ldots, n-1$, approximate the differential equation (8.12) by the difference equation

$$\frac{y_{i-1} - 2y_i + y_{i+1}}{h^2} + P(x_i)\frac{y_{i+1} - y_{i-1}}{2h} + Q(x_i)y_i$$
$$= R(x_i), i = 1, 2, \ldots, n-1, \quad (8.16)$$

or, equivalently, by multiplication and recombination,

$$[2 - hP(x_i)]\, y_{i-1} + \left[-4 + 2h^2 Q(x_i)\right] y_i + [2 + hP(x_i)]\, y_{i+1}$$
$$= 2h^2 R(x_i). \quad (8.17)$$

Step 4. Write down (8.17) in order, for $i = 1, 2, 3, \ldots, n-1$, and insert the known values for y_0, y_n. This yields a tridiagonal linear algebraic equation which satisfies the conditions of Theorem 7.1.

Step 5. Solve the linear algebraic system of Step 4 by either the direct method or by the Newton–Lieberstein method to yield the numerical solution.

Example 8.1 Consider the boundary value problem

$$y'' + \frac{1}{2}(x - 3)y' - y = 0 \quad (8.18)$$

$$y(0) = 11, \quad y(6) = 11. \quad (8.19)$$

On $0 \leq x \leq 6$, $P(x) = \frac{1}{2}(x-3)$, $Q(x) = -1$, $R(x) = 0$. The problem then has a unique solution. Moreover, on $0 \leq x \leq 6$, $\frac{1}{2}|x-3| \leq \frac{3}{2}$, so that one may choose $M = \frac{3}{2}$. Now, divide $0 \leq x \leq 6$ into six equal parts. The grid points are $x_0 = 0$, $x_1 = 1$, $x_2 = 2$, $x_3 = 3$, $x_4 = 4$, $x_5 = 5$, $x_6 = 6$ and the grid size is $h = 1$. Thus, $Mh = \frac{3}{2}$ and (8.15) is satisfied. The differential equation (8.18) is now approximated by the difference equation

$$\frac{y_{i-1} - 2y_i + y_{i+1}}{h^2} + \frac{1}{2}(x_i - 3)\frac{y_{i+1} - y_{i-1}}{2h} - y_i = 0, \quad i = 1, 2, 3, 4, 5 \quad (8.20)$$

8.3 Numerical Solution of Boundary Value Problems Using Difference Equations

which simplifies to

$$(7 - x_i)y_{i-1} - 12y_i + (1 + x_i)y_{i+1} = 0. \tag{8.21}$$

Writing each of (8.21), in order, for $i = 1, 2, 3, 4, 5$, yields

$$(7 - x_1)y_0 - 12y_1 + (1 + x_1)y_2 = 0$$
$$(7 - x_2)y_1 - 12y_2 + (1 + x_2)y_3 = 0$$
$$(7 - x_3)y_2 - 12y_3 + (1 + x_3)y_4 = 0$$
$$(7 - x_4)y_3 - 12y_4 + (1 + x_4)y_5 = 0$$
$$(7 - x_5)y_4 - 12y_5 + (1 + x_5)y_6 = 0.$$

Substitution of the known values for y_0, y_6, and x_1, x_2, x_3, x_4, x_5 into the system yields, then,

$$\begin{aligned}
-12y_1 + 2y_2 &&&&&&&& = -66 \\
5y_1 &- 12y_2 + 3y_3 &&&&&&& = 0 \\
&& 4y_2 &- 12y_3 + 4y_4 &&&& = 0 \\
&&& 3y_3 &- 12y_4 + 5y_5 && = 0 \\
&&&& 2y_4 &- 12y_5 & = -66
\end{aligned}$$

which satisfies the conditions of Theorem 7.1 and has the solution

$$y_1 = 6,\ y_2 = 3,\ y_3 = 2,\ y_4 = 3,\ y_5 = 6. \tag{8.22}$$

This numerical solution agrees exactly with the analytical solution $y = x^2 - 6x + 11$, but this time, not because the differential equation is simple, but because the exact solution has all derivatives of order greater than 2 identically equal to zero. The reason that the exact solution is known is that we started with the elementary function $y = x^2 - 6x + 11$ and then made up the problem (8.18), (8.19), for which it was the solution.

Let us now examine various aspects of the algorithm and show the motivations involved.

The first point to consider is the reason for the assumption $Mh < 2$. To make this clear let us examine (8.21). Observe that, for the tridiagonal system (8.23) which results, finally, the main diagonal elements are the coefficients in (8.21) of y_i, that is, -12, the super diagonal elements which result are the coefficients in (8.21) of y_{i+1}, that is, $(1 + x_i)$, and the sub diagonal elements which result are the coefficients in (8.21) of y_{i-1}, that is, $(7 - x_i)$. From this observation, one sees, with respect to (8.17), that, in general, the main diagonal elements will be

$$-4 + 2h^2 Q(x_i), \tag{8.23}$$

the super diagonal elements will be

$$2 + hP(x_i), \tag{8.24}$$

and the sub diagonal elements will be

$$2 - hP(x_i). \tag{8.25}$$

Now, to satisfy the conditions of Theorem 7.1, we want to be sure first that

$$-4 + 2h^2 Q(x_i) < 0 \tag{8.26}$$
$$2 + hP(x_i) > 0 \tag{8.27}$$
$$2 - hP(x_i) > 0. \tag{8.28}$$

But, by (8.14), $Q(x_i) \leq 0$, so that (8.26) is valid. Moreover, by (8.15),

$$|hP(x)| \leq hM < 2,$$

so that (8.27) and (8.28) are also valid. However, we also want to be sure of diagonal dominance, that is,

$$\left| -4 + 2h^2 Q(x_i) \right| \geq |2 + hP(x_i)| \tag{8.29}$$

$$\left| -4 + 2h^2 Q(x_i) \right| \geq |2 - hP(x_i)|, \tag{8.30}$$

and that inequality is valid for at least one row. But this is also valid. Note first that

$$\left| -4 + 2h^2 Q(x_i) \right| = 4 - 2h^2 Q(x_i) \geq 4.$$

In addition, by (8.27), (8.28),

$$|2 + hP(x_i)| + |2 - hP(x_i)| = 2 + hP(x_i) + 2 - hP(x_i) = 4,$$

so that

$$\left| -4 + 2h^2 Q(x_i) \right| \geq |2 + hP(x_i)| + |2 - hP(x_i)|, \quad i = 1, 2, 3, \ldots, n-1.$$

But, strict inequality is valid for the very first and very last equations, and, hence, all the assumptions of Theorem 7.1 are valid.

8.4
Upwind Differencing

The condition (8.15) has two interesting implications. First, if $P(x) \equiv 0$, so that M can be chosen to be 0, then h can have any positive value at all. However, if $P(x)$ can be large, then M must be large, so that h must be small. Thus, if $M = 2000$ and $a = 0$, $b = 10$, then one must have $h < 0.001$. If one's budget or computer capability does not allow h so small, one can still solve the problem, but with some loss of accuracy. The method one can use is called the *upstream method* and is described in the next example.

Example 8.2 Consider the boundary value problem

$$y'' + 2(2 - x)y' = 2(2 - x) \tag{8.31}$$
$$y(0) = -1, \quad y(6) = 5. \tag{8.32}$$

If one uses the algorithm of Section 8.3 with $h = 1$, then the system of equations which results is

$$\begin{aligned}
-4y_1 + 4y_2 &&&&&= 4 \\
2y_1 - 4y_2 + 2y_3 &&&&&= 0 \\
4y_2 - 4y_3 &&&&&= -4 \\
6y_3 - 4y_4 - 2y_5 &&&&&= -8 \\
8y_4 - 4y_5 &&&&&= 8.
\end{aligned} \tag{8.33}$$

This system does not satisfy the conditions of Theorem 7.1. The problem is that

$$\max |P(x)| = \max |2(2-x)| = 8 \leq M$$

on $0 \leq x \leq 6$ and the condition $hM < 2$ has been violated. Now, it may be that system (8.33) has a unique solution, but we are not sure, *a priori*, that this is correct. Without further analysis, no such conclusion can be reached. Let us then approach (8.31), (8.32) in the following way, still keeping $h = 1$. The grid points are $x_0 = 0$, $x_1 = 1$, $x_2 = 2$, $x_3 = 3$, $x_4 = 4$, $x_5 = 5$, $x_6 = 6$. At each interior grid point approximate (8.31) by

$$\frac{y_{i-1} - 2y_i + y_{i+1}}{h^2} + 2(2 - x_i)v_i = 2(2 - x_i), \quad i = 1, 2, 3, 4, 5 \tag{8.34}$$

where v_i are approximations, as yet unspecified, for y' at x_i, $i = 1, 2, 3, 4, 5$, respectively.

Equations (8.34) are equivalent to

$$\begin{aligned}
-2y_1 + y_2 &+ 2v_1 &&= 3 & (8.35) \\
y_1 - 2y_2 + y_3 &&&= 0 & (8.36) \\
y_2 - 2y_3 + y_4 &- 2v_3 &&= -2 & (8.37) \\
y_3 - 2y_4 + y_5 &- 4v_4 &&= -4 & (8.38) \\
y_4 - 2y_5 &- 6v_5 &&= -11. & (8.39)
\end{aligned}$$

In (8.35)–(8.39), we now choose one of the two formulas (8.3) or (8.4) to approximate y', rather than the slightly more accurate formula (8.5). We will do this carefully, however, so that system (8.35)–(8.39) will satisfy the conditions of Theorem 7.1. Since the coefficient of v_1 is positive in (8.35), let us choose the forward difference approximation (8.3), that is,

$$v_1 = \frac{y_2 - y_1}{h} = y_2 - y_1. \tag{8.40}$$

Substitution of (8.40) into (8.35) yields
$$-4y_1 + 3y_2 = 0, \qquad (8.41)$$
so that the main diagonal element has become more negative than it was in (8.35). Equation (8.36) requires no approximation for v_2. In (8.37), we have to approximate v_3. This time, however, the coefficient of v_3 is negative, so we choose the backward difference approximation
$$v_3 = \frac{y_3 - y_2}{h} = y_3 - y_2, \qquad (8.42)$$
so that (8.37) becomes
$$3y_2 - 4y_3 + y_4 = -2. \qquad (8.43)$$
Similarly, using the backward difference formula (8.4) yields for (8.38) and (8.39), respectively,
$$5y_3 - 6y_4 + y_5 = -4$$
$$7y_4 - 8y_5 = -11.$$

Thus, the system becomes
$$\begin{aligned}
-4y_1 + 3y_2 &= 3 \\
y_1 - 2y_2 + y_3 &= 0 \\
3y_2 - 4y_3 + y_4 &= -2 \\
5y_3 - 6y_4 + y_5 &= -4 \\
7y_4 - 8y_5 &= -11,
\end{aligned}$$
which does satisfy all the conditions of Theorem 7.1 and can be solved by the direct or Newton–Lieberstein methods to yield the approximate solution
$$y_1 = 0, \ y_2 = 1, \ y_3 = 2, \ y_4 = 3, \ y_5 = 4.$$

8.5
Mildly Nonlinear Boundary Value Problems

The most widely studied class of mildly nonlinear boundary value problems is of the form
$$y'' = f(x, y) \qquad (8.44)$$
$$y(a) = \alpha, \quad y(b) = \beta. \qquad (8.45)$$
To ensure that (8.44), (8.45) has a unique solution, we assume that
$$\frac{\partial f(x, y)}{\partial y} \geq 0, \quad a \leq x \leq b, \quad -\infty < y < \infty. \qquad (8.46)$$
When (8.46) is valid, (8.44), (8.45) is called a mildly nonlinear problem.

Note that for the linear equation (8.12), if one assumes y' to be an independent variable, then

$$\frac{\partial}{\partial y}\left[-P(x)y' - Q(x)y + R(x)\right] = -Q(x),$$

so that condition (8.46) implies $-Q(x) \geq 0$, or, $Q(x) \leq 0$, which is (8.14).

The numerical solution of (8.44), (8.45) follows in the manner prescribed in Section 8.4, but with the approximation

$$\frac{y_{i-1} - 2y_i + y_{i+1}}{h^2} = f(x_i, y_i) \tag{8.47}$$

replacing (8.16) and with the resulting nonlinear system being solved by the Newton–Lieberstein method.

Example 8.3 Consider the mildly nonlinear problem

$$y'' = e^y; \quad y(0) = 0, \quad y(1) = 0. \tag{8.48}$$

In this case, $f(x, y) = e^y$, so that $\frac{\partial f(x,y)}{\partial y} = e^y > 0$. Thus, the solution exists and is unique. Numerically, divide $0 \leq x \leq 1$ into five equal parts, so that $h = 0.2$, $x_0 = 0.0$, $x_1 = 0.2$, $x_2 = 0.4$, $x_3 = 0.6$, $x_4 = 0.8$, $x_5 = 1.0$. Let $y(x_i) = y_i$, so that $y_0 = 0$, $y_5 = 0$. To approximate y_1, y_2, y_3, y_4, the differential equation is approximated at each interior grid point by the nonlinear difference equation

$$\frac{y_{i-1} - 2y_i + y_{i+1}}{(0.2)^2} - e^{y_i} = 0, \quad i = 1,2,3,4 \tag{8.49}$$

or, equivalently,

$$25y_{i-1} - 50y_i + 25y_{i+1} - e^{y_i} = 0, \quad i = 1,2,3,4. \tag{8.50}$$

For each of $i = 1,2,3,4$, (8.54) and the zero boundary conditions yield the mildly nonlinear system

$$\begin{array}{rrrrrrl}
-50y_1 & + 25y_2 & & & - e^{y_1} & = 0 \\
25y_1 & - 50y_2 & + 25y_3 & & - e^{y_2} & = 0 \\
& 25y_2 & - 50y_3 & + 25y_4 & - e^{y_3} & = 0 \\
& & 25y_3 & - 50y_4 & - e^{y_4} & = 0.
\end{array}$$

However, this system is exactly the system given in the final example of Section 7.4, but with x replaced by y. Thus, the solution is

$$y_1 = -0.0731, \quad y_2 = -0.1089, \quad y_3 = -0.1089, \quad y_4 = -0.0731.$$

8.6
Theoretical Support*

The theory for the numerical solution of linear and nonlinear boundary value problems is more complex than that for initial value problems (Bers (1953); Greenspan and Casulli (1988); Keller (1968)). In this section, then, we will concentrate only on linear boundary value problems of a special type.

Consider the linear boundary value problem

$$y'' + P(x)y' + Q(x)y = R(x) \tag{8.51}$$
$$y(a) = \alpha, \quad y(b) = \beta \tag{8.52}$$

but with (8.14) replaced by the stronger condition that for some $\epsilon > 0$,

$$Q(x) < -\epsilon. \tag{8.53}$$

We assume that (8.51)–(8.53) has been solved by the method of Section 8.3.

For any function $v(x, y)$ which is continuous on $a < x < b$ and which has continuous derivatives on $a < x < b$, the operators L and L_h are defined at each interior grid point x_i, $i = 1, 2, 3, \ldots, n-1$, by

$$Lv(x_i) = v''(x_i) + P(x_i)v'(x_i) + Q(x_i)v_i \tag{8.54}$$

$$L_h v(x_i) = \frac{v_{i-1} - 2v_i + v_{i+1}}{h^2} + P(x_i)\frac{v_{i+1} - v_{i-1}}{2h} + Q(x_i)v_i. \tag{8.55}$$

Finally, let positive bounds M, N_1, N_2 be defined by

$$M \geq |P(x)|, \quad a < x < b \tag{8.56}$$
$$N_1 \geq -Q(x) \geq N_2 > 0, \quad a < x < b \tag{8.57}$$

We now have the following lemma.

Lemma 8.1 *Let v be a discrete function defined on $a = x_0, x_1, x_2, \ldots, x_n = b$. Let h satisfy $h < 2/M$. Define C by*

$$C = \max\left[1, \frac{1}{N_2}\right].$$

Then,

$$|v(x_i)| \leq C\left\{\max\left[|v_0|, |v_n|\right] + \max_{1 \leq j \leq n-1}|L_h v(x_j)|\right\} \tag{8.58}$$

Proof. If $\max |v(x_i)|$ occurs for $i = 0$ or $i = n$, then (8.58) is valid since $C \geq 1$. Hence we proceed under the assumption that $\max |v(x_i)|$ is attained for i equal to one of $1, 2, 3, \ldots, n-1$.

For definiteness, assume that the algebraic equations in our numerical method are written in the form

$$\frac{1}{2}h^2 L_h v(x_i) = \frac{1}{2}h^2 R(x_i),$$

so that from (8.55), for $i = 1, 2, 3, \ldots, n-1$,

$$\frac{1}{2}h^2 L_h v(x_i) = \left[\frac{1}{2} - \frac{1}{4}hP(x_i)\right] v_{i-1} + \left[-1 + \frac{1}{2}h^2 Q(x_i)\right] v_i \qquad (8.59)$$
$$+ \left[\frac{1}{2} + \frac{1}{4}hP(x_i)\right] v_{i+1}.$$

Thus the coefficient matrix of the resulting tridiagonal system will have main diagonal elements

$$a_{ii} = -1 + \frac{1}{2}h^2 Q(x_i) < -1, \quad i = 1, 2, \ldots, n \qquad (8.60)$$

will have sub diagonal elements

$$a_{i,i-1} = \frac{1}{2} - \frac{1}{4}hP(x_i) > 0, \quad i = 2, 3, \ldots, n \qquad (8.61)$$

and will have super diagonal elements

$$a_{i+1,i} = \frac{1}{2} + \frac{1}{4}hP(x_i) > 0, \quad i = 1, 2, \ldots, n-1. \qquad (8.62)$$

For $i = 1, 2, \ldots, n-1$, consider now (8.59) in the equivalent form

$$-\left[1 - \frac{1}{2}h^2 Q(x_i)\right] v_i = -\left[\frac{1}{2} - \frac{1}{4}hP(x_i)\right] v_{i-1} - \left[\frac{1}{2} + \frac{1}{4}hP(x_i)\right] v_{i+1}$$
$$+ \frac{1}{2}h^2 L_h v(x_i).$$

Taking absolute values in this last relation implies readily, with the aid of (8.57) and (8.60)–(8.62) that, for $i = 1, 2, 3, \ldots, n-1$,

$$\left[1 + \frac{1}{2}h^2 N_2\right] \cdot |v_i| \leq \max_{j=1,2,\ldots,n-1} |v_j| + \frac{1}{2}h^2 \max_{j=1,2,\ldots,n-1} |L_h v(x_i)|$$

Since the right side of this inequality is independent of i, it follows that if one chooses that value of i for which $|v_i|$ is a maximum, then the inequality yields

$$N_2 \max_i |v_i| \leq \max_i |L_h v(x_i)|, \quad i = 1, 2, 3, \ldots, n-1.$$

Hence, for $i = 1, 2, 3, \ldots, n-1$,

$$\max_i |v_i| \leq C \max_i |L_h v(x_i)| \leq C\{\max[|v_0|, |v_n|] + \max_i |L_h v(x_i)|\},$$

and the lemma is proved. □

With the aid of Lemma 8.1, we now prove a major convergence theorem.

Theorem 8.1 *Let $y(x)$ be the exact solution of boundary value problem (8.51)–(8.53). Let u_i, $i = 1, 2, 3, \ldots, n-1$ be the numerical solution on x_i, $i = 1, 2, 3, \ldots, n-1$ generated by the numerical method of Section 8.3. Under the assumptions of Lemma 8.1, if $y(x)$ has continuous derivatives up to and including order 4 on $a \leq x \leq b$, then*

$$|u_i - y(x_i)| \leq \frac{1}{12} Ch^2 (M_4 + 2MM_3), \quad i = 0, 1, 2, 3, \ldots, n, \tag{8.63}$$

where

$$M_r = \max_{a \leq x \leq b} \left| \frac{d^r y}{dx^r} \right|, \quad r = 3, 4. \tag{8.64}$$

Proof. The validity of (8.63) for $i = 0$, $i = n$ is a direct consequence of the numerical method, since $u_0 = y(x_0) = \alpha$, $u_n = y(x_n) = \beta$. Hence, we need only consider $i = 1, 2, 3, \ldots, n-1$. Set

$$v_i = u_i - y(x_i)$$

into (8.58). Since $v_0 = v_n = 0$, one has

$$|u_i - y(x_i)| \leq C \max_i |L_h [u_i - y(x_i)]|, \quad i = 1, 2, 3, \ldots, n-1. \tag{8.65}$$

But,

$$\begin{aligned} L_h [u_i - y(x_i)] &= L_h u_i - L_h y(x_i) + Ly(x_i) - Ly(x_i) \\ &= R(x_i) - L_h y(x_i) + Ly(x_i) - R(x_i) \\ &= y''(x_i) + P(x_i) y'(x_i) + Q(x_i) y_i \\ &\quad - \frac{y_{i-1} - 2y_i + y_{i+1}}{h^2} - P(x_i) \frac{y_{i+1} - y_{i-1}}{2h} - Q(x_i) y_i, \end{aligned}$$

which, by insertion of Taylor expansions through fourth-order for y_{i-1} and y_{i+1} into $y_{i-1} - 2y_i + y_{i+1}$ and through third-order for y_{i-1} and y_{i+1} into $y_{i+1} - y_{i-1}$ implies that

$$\begin{aligned} L_h [u_i - y(x_i)] = &-\frac{1}{24} h^2 \frac{d^4 y(\delta_1)}{dx^4} - \frac{1}{24} h^2 \frac{d^4 y(\delta_2)}{dx^4} \\ &- \frac{1}{12} h^2 P(x_i) \frac{d^3 y(\delta_3)}{dx^3} - \frac{1}{12} h^2 P(x_i) \frac{d^3 y(\delta_4)}{dx^3}. \end{aligned}$$

Thus

$$|L_h [u_i - y(x_i)]| \leq \frac{1}{12} h^2 M_4 + \frac{1}{6} h^2 M M_3,$$

which, upon substitution into (8.65) yields (8.63) and the theorem is proved. □

The inequality (8.63) is a nonconstructive error bound because we have no apparent means to estimate M_3 and M_4. However, it does allow us to conclude that, for the class of problems to which the theorem is applicable, one has convergence of the numerical to the exact solution as the grid size converges to zero, that is, $\lim_{h \to 0} u_i = y(x_i)$.

8.7
Application – Approximation of Airy Functions

The mathematician-astronomer G. B. Airy made extensive studies of the equation

$$y'' - xy = 0, \quad -\infty < x < \infty \tag{8.66}$$

in connection with the theory of diffraction (Boyce and diPrima (1986), Simmons (1972)). Solutions of (8.66) are called Airy functions and these can be generated by power series. However, suppose in addition to (8.66) one is given the two boundary conditions

$$y(0) = 2, \quad y(6) = 1. \tag{8.67}$$

Then, only one of the arbitrary constants in the general solution can be determined exactly. Indeed, if one wishes merely to know the solution at $x = 4$, this also cannot be determined exactly from the series solution. However, the approximation of Airy functions at various points can be determined by the algorithm of Section 8.4. Consider, for example, problem (8.66), (8.67). Divide the interval $0 \le x \le 6$ into five equal parts so that $x_0 = 0.0$, $x_1 = 1.0$, $x_2 = 2.0$, $x_3 = 3.0$, $x_4 = 4.0$, $x_5 = 5.0$, $x_6 = 6.0$, $h = 1.0$. Then, $y_0 = 2$, $y_6 = 1$. At each interior point, (8.66) is approximated by

$$y_{i-1} - 2y_i + y_{i+1} - x_i y_i = 0,$$

or,

$$y_{i-1} - (2 + x_i)y_i + y_{i+1} = 0, \quad i = 1, 2, 3, 4, 5.$$

The resulting tridiagonal system is

$$
\begin{aligned}
-3y_1 + y_2 &= -2 \\
y_1 - 4y_2 + y_3 &= 0 \\
y_2 - 5y_3 + y_4 &= 0 \\
y_3 - 6y_4 + y_5 &= 0 \\
y_4 - 7y_5 &= -1,
\end{aligned}
$$

the solution of which is

$$y_1 = 0.7314, \quad y_2 = 0.1942, \quad y_3 = 0.0453, \quad y_4 = 0.0321, \quad y_5 = 0.1474,$$

thus yielding approximations of the Airy function. For improved accuracy, one need only decrease h.

8.8
Exercises

8.1 Using (8.16), determine what choices of h will ensure that the numerical solution of

$$y'' + 4(\sin x)y' - 4(\cos x)y = -\sin x, \quad y(0) = 0, \quad y\left(\frac{1}{2}\pi\right) = 1$$

exists and is unique.

8.2 Using $h = \frac{1}{100}\pi$ with (8.16), find the numerical solution of the boundary value problem given in Exercise 1. Compare your results with the exact solution $y = \sin x$.

8.3 Using $h = 0.01$, find a numerical solution for each of the following boundary value problems and compare your results with the exact solution:

(a) $y'' + y' - y = 1 - x$, $\quad y(0) = 0$, $\quad y(1) = 1$ $\quad$ (Exact: $y = x$)
(b) $y'' + xy' - 2y = 2$, $\quad y(-1) = 1$, $\quad y(1) = 1$ $\quad$ (Exact: $y = x^2$)
(c) $y'' + xy' - 3y = 6x$, $\quad y(-1) = -1$, $\quad y(1) = 1$ $\quad$ (Exact: $y = x^3$).

8.4 Using $h = 0.01$, find a numerical solution for each of the following boundary value problems:

(a) $y'' = e^y$, $\quad y(0) = 0$, $\quad y(1) = 1$
(b) $y'' = y^3$, $\quad y(0) = 0$, $\quad y(1) = 1$
(c) $y'' = 2 + e^y$, $\quad y(0) = 0$, $\quad y(1) = 0$
(d) $y'' + y' = y^3$, $\quad y(0) = 0$, $\quad y(1) = 1$.

8.5 Using $h = 0.01$, find a numerical solution for each of the following boundary value problems and compare your results with the exact solution:

(a) $y'' = \frac{1}{2}(1 + x + y)^3$, $y(0) = 0$, $y(1) = 0$ $\left(\text{Exact: } y = -x - 1 + \frac{2}{2-x}\right)$
(b) $(0.01)y'' + y' = 0$, $\quad y(0) = 0$, $y(\infty) = 1$ (Exact: $y = 1 - e^{-100x}$).

8.6 Find an approximate solution of the boundary value problem

$$y'''' + y'' = 2, \; y(0) = y'(0) = 0, \; y(1) = 1, \; y'(1) = 2.$$

Compare your results with the exact solution $y = x^2$.

8.7 Show that the boundary value problem

$$y'' = e^y, \quad y(0) = p, \quad y(1) = q$$

has the solution

$$y = \log c_1 - 2\log\left\{\cos\left[\left(\frac{1}{2}c_1\right)^{\frac{1}{2}} x + c_2\right]\right\}$$

where c_1, c_2 are solutions of

$$p = \log c_1 - 2\log(\cos c_2)$$
$$q = \log c_1 - 2\log\left\{\cos\left[\left(\frac{1}{2}c_1\right)^{\frac{1}{2}} x + c_2\right]\right\}.$$

9
Special Relativistic Motion

9.1
Introduction

Deterministic particle motion can be studied also in a relativistic context. We will do this in the present chapter. However, only pertinent concepts from special relativity will be defined and applied. In addition, we will concentrate on the fundamental problem of motion in one space dimension.

In Newtonian mechanics one assumes that in making observations the speed of light is infinite. Thus, in Newtonian mechanics one assumes that an observed event is seen at the exact time it occurs. In fact, the speed of light is not infinite but is, approximately, 186 000 mile/s. This constant will be denoted by c. That the time involved in making an observation can be significant is apparent in the following two cases:

(a) when observing very distant objects, like galaxies, and

(b) when observing objects moving close to the speed of light, like

an electron which has been accelerated to $0.95c$.

The subject in which the finite speed of light is taken into account is called relativity. Special relativity uses the Euclidean formula to measure the distance between two points, while general relativity uses a more general metric. We will deal only with special relativity.

Special relativity has a fundamental restriction in its application which will now be described. In order to understand this restriction let us recall the definition of an N-body problem, which was formulated as follows. In cgs units and for $i = 1, 2, \ldots, N$, let P_i of mass m_i be at $\vec{r}_i = (x_i, y_i, z_i)$, have velocity $\vec{v}_i = (v_{i,x}, v_{i,y}, v_{i,z})$, and have acceleration $\vec{a}_i = (a_{i,x}, a_{i,y}, a_{i,z})$ at time $t \geq 0$. Let the positive distance between P_i and P_j, $i \neq j$, be $r_{ij} = r_{ji} \neq 0$. Let the force on P_i due to P_j be $\vec{F}_{ij} = \vec{F}_{ij}(r_{ij})$. Also, assume that the force $\vec{F}_{ji}$ on P_j due to

Numerical Solution of Ordinary Differential Equations for Classical, Relativistic and Nano Systems. Donald Greenspan
Copyright © 2006 WILEY-VCH Verlag GmbH & Co. KGaA, Weinheim
ISBN: 3-527-40610-7

P_i satisfies $\vec{F}_{ji} = -\vec{F}_{ij}$. Then, given the initial positions and velocities of all the $P_i, i = 1, 2, 3, \ldots, N$, the general N-body problem is to determine the motion of the system if each P_i interacts with all the other P_j's in the system.

The statement of the N-body problem has neglected a very subtle aspect in the formulation, which is that the forces $\vec{F}_{ij}$ and $\vec{F}_{ji}$ act at the same time, that is they act simultaneously. Let us show by a simple example given by Einstein that simultaneity cannot exist in special relativity. The implication of the example is that in special relativity one must restrict attention to the motion of a single body. One cannot, for example, use special relativity to simulate the motions of two or more planets in the solar system.

Einstein's example proceeds as follows. Three men A, O, and B are riding on a train which moves at the constant positive speed V on a straight track. A is in the front of the train, B is at the rear and O is exactly in the middle. A fourth man O' is standing beside the rails. At the very instant O passes O', it happens that two flash bulb signals coming from A and B reach O and O'. The problem is to determine which of A or B emitted his signal first. Since O is on the train with A and B and knows that he is exactly in the middle between them, he concludes readily that the signals were sent at the same time. But O' reasons differently as follows. The two flashes arrived when the middle of the train was passing him. Therefore, the flashes were emitted before the middle of the train reached him, since the speed of light is finite. But, just before the middle of the train reached him, A was nearer to him than was B. Thus the light from B had farther to travel to reach him and took a longer time. But both reached him at the same time, so that B had to emit his flash before A did.

The example demonstrates that if one allows the concept of simultaneity in special relativity, then contrary conclusions can be reached. Hence, simultaneity is non admissible in the theory.

Let us proceed then to fundamental concepts.

9.2
Inertial Frames

Consider two x axes, which at some time $t = 0$ are *coincident* and *identical*. Denote one by X and the second by X'. At the origin of each assume there are identical observers, each with identical clocks synchronized at $t = 0$. Assume finally that the X' axis is in motion relative to the X axis with constant speed u. The X and X' axes are called inertial frames, or systems, the X axis being called the *lab*, or *laboratory*, frame and the X' axis being called the *rocket* frame. The term *rocket frame* is used to denote the importance of u being a large constant.

We assume the axiom of continuity for inertial frames, that is: all the laws of physics are the same in every inertial reference frame, and this includes the invariance of all physical constants, including the speed of light.

Fig. 9.1 Two inertial frames.

9.3
The Lorentz Transformation

Consider now two inertial reference frames X and X' at some time $t > 0$, as shown in Figure 9.1. Suppose the observers at O and O' both observe an event, like an exploding star P. Let P occur in the lab frame at position x and at time t, while in the rocket frame it occurs at position x' and at time t'. Then (x, t) and (x', t') are called events because they incorporate knowledge of both position and time of occurrence. Taking into account the speed of light in making observations, H. A. Lorentz proved that the relationships between (x, t) and (x', t') are

$$x' = \frac{c(x - ut)}{(c^2 - u^2)^{\frac{1}{2}}}, \quad t' = \frac{c^2 t - ux}{c(c^2 - u^2)^{\frac{1}{2}}} \tag{9.1a}$$

or, equivalently,

$$x = \frac{c(x' + ut')}{(c^2 - u^2)^{\frac{1}{2}}}, \quad t = \frac{c^2 t' + ux'}{c(c^2 - u^2)^{\frac{1}{2}}}. \tag{9.1b}$$

Equations (9.1) are called the Lorentz transformation.
To avoid singularities, we assume that $|u| < c$.

9.4
Rod Contraction and Time Dilation

The Lorentz formulas have interesting physical implications, two of which will be discussed now. The first is called the contraction of a moving rod and is described as follows.

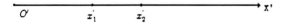

Fig. 9.2 Rod at rest in the rocket frame.

Consider a thin rod which lies in the lab frame. Then, the rod's length is measured to be greatest when the rod is at rest relative to the lab frame observer. When it moves with speed u relative to the observer its measured length is contracted by the factor $\sqrt{1 - \frac{u^2}{c^2}}$. To see this, let the rod lie at rest along the X' axis in the rocket frame, as shown in Figure 9.2. Let its end points be x'_2 and x'_1, with $x'_2 > x'_1$, so that its length is $x'_2 - x'_1$. Now, at time t, let the rod's length be measured by the lab frame observer. Then the end point coordinates x_2, x_1 as observed in the lab frame are related to x'_2, x'_1 by

$$x'_2 = \frac{c(x_2 - ut)}{(c^2 - u^2)^{\frac{1}{2}}}, \quad x'_1 = \frac{c(x_1 - ut)}{(c^2 - u^2)^{\frac{1}{2}}}. \quad (9.2)$$

Thus,

$$x'_2 - x'_1 = \frac{c(x_2 - x_1)}{(c^2 - u^2)^{\frac{1}{2}}},$$

or,

$$x_2 - x_1 = (x'_2 - x'_1)\left(\frac{c^2 - u^2}{c^2}\right)^{\frac{1}{2}} = (x'_2 - x'_1)\sqrt{1 - \frac{u^2}{c^2}}. \quad (9.3)$$

However,

$$\sqrt{1 - \frac{u^2}{c^2}} < 1, \quad (9.4)$$

which proves the contention, if one also observes that $\sqrt{1 - \frac{u^2}{c^2}} = 1$ if and only if $u = 0$.

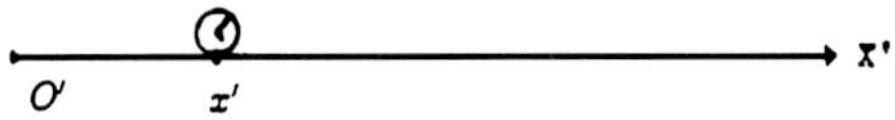

Fig. 9.3 Clock at rest in the rocket frame.

Next observe that a clock is measured in the lab frame to go at its fastest rate when it is at rest relative to the lab frame observer. When the clock moves with a speed u relative to the lab observer, its rate is measured to have slowed down by the factor $\sqrt{1 - \frac{u^2}{c^2}}$. If its rate of ticking slows down, time is dilated, that is, for example, the clock takes longer to measure an hour. To prove this, consider a clock which is at rest at a position x' in the rocket frame, as shown in Figure 9.3. Consider two successive readings in the rocket frame, say at t'_1 and t'_2.

The time interval in the rocket is $t'_2 - t'_1$. To the lab observer, these times are recorded as t_1 and t_2, where

$$t_1 = \frac{c^2 t'_1 + ux'}{c(c^2 - u^2)^{\frac{1}{2}}}, \quad t_2 = \frac{c^2 t'_2 + ux'}{c(c^2 - u^2)^{\frac{1}{2}}}. \quad (9.5)$$

However,

$$t_2 - t_1 = \frac{c^2(t'_2 - t'_1)}{c(c^2 - u^2)^{\frac{1}{2}}} = \frac{(t'_2 - t'_1)}{\sqrt{1 - \frac{u^2}{c^2}}}. \quad (9.6)$$

But,
$$\frac{1}{\sqrt{1-\frac{u^2}{c^2}}} > 1, \qquad (9.7)$$
from which the assertion follows.

9.5
Relativistic Particle Motion

Consider now a particle P in motion in the lab frame. Then its velocity v and acceleration a are defined in the usual way:

$$v = \frac{dx}{dt}, \quad a = \frac{dv}{dt}. \qquad (9.8)$$

By the axiom of continuity, one must have in the rocket

$$v' = \frac{dx'}{dt'}, \quad a' = \frac{dv'}{dt'}. \qquad (9.9)$$

With regard to (9.8) and (9.9), we assume $|v| < c$, $|v'| < c$.

To relate v and v', we have from (9.1a)

$$v' = \frac{dx'}{dt'} = \frac{c(dx - u\,dt)}{(c^2 - u^2)^{\frac{1}{2}}} \div \frac{c^2 dt - u\,dx}{c(c^2 - u^2)^{\frac{1}{2}}} = \frac{c^2(dx - u\,dt)}{c^2 dt - u\,dx},$$

so that

$$v' = \frac{c^2(v - u)}{c^2 - uv}. \qquad (9.10)$$

Equivalently,

$$v = \frac{c^2(v' + u)}{c^2 + uv'}. \qquad (9.11)$$

Similarly, the relationship between a and a' is found to be

$$a' = \frac{c^3(c^2 - u^2)^{\frac{3}{2}}}{(c^2 - uv)^3} a \qquad (9.12)$$

or, equivalently,

$$a = \frac{c^3(c^2 - u^2)^{\frac{3}{2}}}{(c^2 + uv')^3} a'. \qquad (9.13)$$

9.6
Covariance

Again, by covariance we mean that the structure of the dynamical equations associated with a physical formulation is invariant under fundamental coordinate transformations. In special relativity, this means under the Lorentz transformation.

Around 1900, it was shown that Newton's dynamical equation was not invariant under the Lorentz transformation, whereas Maxwell's equations, that is, the equations of electromagnetics, were. The question arose, then, as to what dynamical equation for particle motion was covariant under the Lorentz transformation, and Einstein showed that if one assumes that mass varies with speed, then with only a slight modification of Newton's equation the result was a covariant dynamical equation. Let us prove this result first, since it is essential for an understanding of later discussions.

Theorem 9.1 *Let a particle P be in motion along the X axis in the lab frame and along the X' axis in the rocket frame. In the lab frame, let the mass m of P be given by*

$$m = \frac{cm_0}{(c^2 - v^2)^{\frac{1}{2}}}, \qquad (9.14)$$

where m_0 is a positive constant called the rest mass of P and v is the speed of P in the lab. In the rocket, let the mass m' of P be given by

$$m' = \frac{cm_0}{[c^2 - (v')^2]^{\frac{1}{2}}}, \qquad (9.15)$$

where m_0 is the same positive constant as in (9.14) and v' is the speed of P in the rocket. Let a force F be applied to P in the lab. In the rocket coordinates, denote the force by F', so that

$$F = F'.$$

Then, if in the lab, the equation of motion of P is given by

$$F = \frac{d}{dt}(mv), \qquad (9.16)$$

it follows that the equation of motion in the rocket is

$$F' = \frac{d}{dt'}(m'v'). \qquad (9.17)$$

Proof. From (9.14) and (9.16)

$$F = v\frac{dm}{dt} + m\frac{dv}{dt}$$

$$= v\left[\frac{-\frac{1}{2}(cm_0)}{(c^2 - v^2)^{\frac{3}{2}}}(-2va)\right] + ma$$

$$= \frac{v^2 ma}{c^2 - v^2} + ma,$$

so that
$$F = \left(\frac{c^2}{c^2 - v^2}\right) ma. \tag{9.18}$$

From (9.15) and (9.17), then, we must have
$$F' = \left[\frac{c^2}{c^2 - (v')^2}\right] m'a'. \tag{9.19}$$

Since $F = F'$, the proof will follow if we can establish the identity
$$\left(\frac{c^2}{c^2 - v^2}\right) ma \equiv \left[\frac{c^2}{c^2 - (v')^2}\right] m'a'.$$

or, equivalently,
$$\frac{ma}{c^2 - v^2} \equiv \frac{m'a'}{c^2 - (v')^2}. \tag{9.20}$$

However, substitution of (9.10), (9.12) and (9.15) into the right side-hand of (9.20) yields, quite remarkably, that the identity is valid and the theorem is proved. □

9.7
Particle Motion

Let us consider now a particle P on the X axis in the lab frame. Assume also that the force F on P is one whose magnitude depends only on the x coordinate of P. Then, let
$$F = f(x). \tag{9.21}$$

Assume that initially, that is, at time $t = 0$, P is at x_0 and has speed v_0. Then the equation of motion of P in the lab frame is
$$\frac{d}{dt}(mv) = f(x). \tag{9.22}$$

From (9.18)
$$\left(\frac{c^2}{c^2 - v^2}\right) ma = f(x), \tag{9.23}$$

or,
$$c^2 m\ddot{x} = f(x)(c^2 - \dot{x}^2)$$

From (9.14), this can be reduced to
$$c^3 m_0 \ddot{x} = f(x)(c^2 - \dot{x}^2)^{\frac{3}{2}},$$

so that, finally,
$$\ddot{x} - \frac{f(x)}{c^3 m_0}(c^2 - \dot{x}^2)^{\frac{3}{2}} = 0 \tag{9.24}$$

is the differential equation one has to solve in the lab frame, given the initial data

$$x(0) = x_0, \quad \dot{x}(0) = v_0. \tag{9.25}$$

In general, (9.23) cannot be solved in closed form, so that the observer in the lab frame must now introduce a computer to approximate the solution. However, the observer in the rocket frame also observes the motion of P, but in his coordinate system. His equation and initial conditions are found by applying (9.1b), (9.11) and (9.13) to (9.24) and (9.25). Thus, he too will be confronted with a differential equations problem which requires computational methodology, so a computer identical to that in the lab is introduced also into the rocket.

A fundamental problem in preserving the physics of special relativity then arises as to how the two observers should approximate the solutions of their initial value problems. Their differential equations are covariant under the Lorentz transformation. To preserve the physics, if they use difference approximations, then the difference approximations of these differential equations should also be covariant under the Lorentz transformation. We will show how to do this next and, indeed, show later that their numerical results are also related by the Lorentz transformation.

9.8
Numerical Methodology

In the lab, let $\Delta t > 0$ and $t_k = k\Delta t$. At time t_k, let P be at x_k in the lab. Then, in the rocket, P will be at x'_k at time t'_k, where

$$x'_k = \frac{c(x_k - ut_k)}{(c^2 - u^2)^{\frac{1}{2}}}, \quad t'_k = \frac{c^2 t_k - ux_k}{c(c^2 - u^2)^{\frac{1}{2}}} \tag{9.26}$$

or, equivalently,

$$x_k = \frac{c(x'_k + ut'_k)}{(c^2 - u^2)^{\frac{1}{2}}}, \quad t_k = \frac{c^2 t'_k + ux'_k}{c(c^2 - u^2)^{\frac{1}{2}}}. \tag{9.27}$$

The formulas (9.26) and (9.27) are valid because they are merely special cases of (9.1), that is, they result from the particular choices $x = x_k$, $t = t_k$.

The concepts of velocity and acceleration are now approximated by the following formulas. At t_k in the lab, let

$$v_k = \frac{\Delta x_k}{\Delta t_k} = \frac{x_{k+1} - x_k}{t_{k+1} - t_k}, \quad a_k = \frac{\Delta v_k}{\Delta t_k} = \frac{v_{k+1} - v_k}{t_{k+1} - t_k}. \tag{9.28}$$

9.8 Numerical Methodology

At t'_k in the rocket, let

$$v'_k = \frac{\Delta x'_k}{\Delta t'_k} = \frac{x'_{k+1} - x'_k}{t'_{k+1} - t'_k}, \; a'_k = \frac{\Delta v'_k}{\Delta t'_k} = \frac{v'_{k+1} - v'_k}{t'_{k+1} - t'_k}. \tag{9.29}$$

Then, corresponding to (9.10)–(9.13), one has by direct substitution

$$v'_k = \frac{c^2(v_k - u)}{c^2 - uv_k}, \; v_k = \frac{c^2(v'_k + u)}{c^2 + uv'_k}. \tag{9.30}$$

Similarly, the relationship between a and a' is found to be

$$a'_k = \frac{c^3(c^2 - u^2)^{\frac{3}{2}}}{(c^2 - uv_k)^2(c^2 - uv_{k+1})} a_k, \; a_k = \frac{c^3(c^2 - u^2)^{\frac{3}{2}}}{(c^2 + uv'_k)^2(c^2 + uv'_{k+1})} a'_k. \tag{9.31}$$

Of course, in the limit, (9.30) and (9.31) converge appropriately to (9.10)–(9.13). Our problem now is one of choosing an approximation to

$$F = \frac{d}{dt}(mv) \tag{9.32}$$

in the lab which will transform covariantly into the rocket. The clue for this choice comes from (9.18), which is equivalent to (9.32). What we choose at t_k is the approximation

$$F_k = \frac{m(t_k)c^2}{\left[(c^2 - v_k^2)(c^2 - v_{k+1}^2)\right]^{\frac{1}{2}}} a_k, \; m(t_k) = \frac{cm_0}{\sqrt{c^2 - v_k^2}}. \tag{9.33}$$

Note first that, again, in the limit, (9.33) converges to (9.18). What we must prove is that if $F_k = F'_k$ and if in the rocket

$$m'(t'_k) = \frac{cm_0}{\sqrt{c^2 - (v'_k)^2}} \tag{9.34}$$

then, in the rocket,

$$F'_k = \frac{m'(t'_k)c^2}{\left[(c^2 - (v'_k)^2)(c^2 - (v'_{k+1})^2)\right]^{\frac{1}{2}}} a'_k. \tag{9.35}$$

Theorem 9.2 *If $F_k = F'_k$, then (9.33), (9.34) imply (9.35).*

Proof. The proof is entirely analogous to that of Theorem 9.1 but uses (9.26)–(9.31). □

We will show now how, in the lab, (9.33) is applied to generate the numerical solution of an initial value oscillator problem.

For simplicity in the discussion which follows, we will use the customary normalization constants $m_0 = c = 1$. Then recursion formulas for the motion of an oscillator in the lab, from (9.28) and (9.33) are

$$x_{k+1} = x_k + (t_{k+1} - t_k)v_k \tag{9.36}$$

$$v_{k+1} = v_k + (t_{k+1} - t_k)(1 - v_k^2)(1 - v_{k+1}^2)^{\frac{1}{2}} F_k. \tag{9.37}$$

Without loss of generality, we consider only $k = 0$, since each recursive step is essentially the same except for subscript. Hence,

$$x_1 = x_0 + (t_1 - t_0)v_0 \tag{9.38}$$

$$v_1 = v_0 + (t_1 - t_0)(1 - v_0^2)(1 - v_1^2)^{\frac{1}{2}} F_0. \tag{9.39}$$

The problem in applying (9.38) and (9.39) is that (9.39) is implicit in v_1. Let us show, however, that (9.31) can be solved explicitly and uniquely for v_1. Set

$$A = (t_1 - t_0)(1 - v_0^2)F_0.$$

Then, (9.39) reduces to

$$v_1 - v_0 = A(1 - v_1^2)^{\frac{1}{2}}. \tag{9.40}$$

Now, $1 - v_0^2 > 0$, $1 - v_1^2 > 0$, $1 - u^2 > 0$ by the assumption that no speed is greater than c, which has been normalized to unity. Thus, if $F_0 = 0$, then $A = 0$. Moreover, if $F_0 > 0$ then $A > 0$, while if $F_0 < 0$, then $A < 0$. Thus, A has the same sign as F_0. Now, if $A = 0$, then $v_1 = v_0$, and so v_1 is unique. Assume then that $A > 0$. Then, squaring both sides of (9.40) yields

$$v_1^2 - 2v_1 v_0 + v_0^2 = A^2(1 - v_1^2) \tag{9.41}$$

so that

$$v_1^2(1 + A^2) - 2v_1 v_0 + v_0^2 - A^2 = 0. \tag{9.42}$$

The solutions to this quadratic are

$$v_1 = \frac{v_0 \pm A\sqrt{1 - v_0^2 + A^2}}{1 + A^2}. \tag{9.43}$$

However, substitution of (9.43) into (9.40) reveals that only

$$v_1 = \frac{v_0 + A\sqrt{1 - v_0^2 + A^2}}{1 + A^2} \tag{9.44}$$

is a root of (9.43). The same result follows in the case $A < 0$. Thus, (9.36)–(9.37) can be replaced by the explicit formulas

$$x_{k+1} = x_k + (t_{k+1} - t_k)v_k \tag{9.45}$$

$$v_{k+1} = \frac{v_k + (t_{k+1} - t_k)(1 - v_k^2)F_k\sqrt{1 - v_k^2 + (t_{k+1} - t_k)^2(1 - v_k^2)^2 F_k^2}}{1 + (t_{k+1} - t_k)^2(1 - v_k^2)^2 F_k^2}. \tag{9.46}$$

9.9
Relativistic Harmonic Oscillation

For the usual definition of a harmonic oscillator in Newtonian mechanics, one assumes that $f(x)$ in (9.21) is given by $f(x) = -K^2 x$, where K is a nonzero constant. We will assume that this same choice of $f(x)$ in (9.22) defines a relativistic harmonic oscillator and examine its motion in the lab frame for given initial data.

We now set $K = 1$, in addition to $m_0 = c = 1$. Then (9.24) reduces to

$$\ddot{x} - x(c^2 - \dot{x}^2)^{\frac{3}{2}} = 0. \tag{9.47}$$

To generate a numerical solution, observe now that for $k = 0, 1, 2, \ldots$,

$$x_{k+1} = x_k + (\Delta t_k)v_k \tag{9.48}$$

$$v_{k+1} = \frac{v_k - (\Delta t_k)x_k(1 - v_k^2)^{\frac{3}{2}}\sqrt{1 + x_k^2(\Delta t_k)^2(1 - v_k^2)}}{1 + (\Delta t_k)^2(1 - v_k^2)^2 x_k^2}. \tag{9.49}$$

Let us assume now that $x(0) = x_0 = 0$ and examine the results for various values of v_0 in the range $0 < v_0 < 1$. In particular this is done for 30 000 time steps with $\Delta t_k = \Delta t = 0.0001$ for each of the cases $v_0 = 0.001, 0.01, 0.05, 0.1, 0.3, 0.5, 0.7, 0.9$.

Figure 9.4 shows the amplitude and period of the first complete oscillation for the case $v_0 = 0.001$. For such a relatively low velocity, the oscillator should behave like a Newtonian oscillator, and, indeed, this is the case with the amplitude being 0.001 and, to two decimal places, the period being 6.28 ($\sim 2\pi$). Subsequent motion of this oscillator continues to show almost no change in amplitude or period. At the other extreme, Figure 9.5 shows the motion for $v_0 = 0.9$, which is relatively close to the speed of light. To two decimal places, the amplitude of the first oscillation is 1.61 while the period is 8.88. These results are distinctly non-Newtonian, and to 30 000 steps these results remain constant to two decimal places but do show small increments in the third decimal place. Finally, in Figure 9.6 is shown how the amplitude of the relativistic harmonic oscillator deviates from that of the Newtonian harmonic oscillator with increasing v_0.

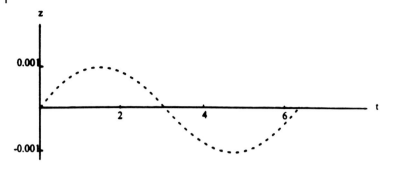

Fig. 9.4 Amplitude and period for relatively low velocity.

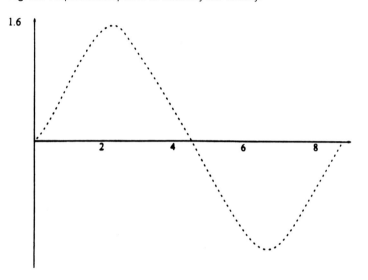

Fig. 9.5 Amplitude and period for velocity near the speed of light.

9.10
Computational Covariance

The discussions in Sections 9.8 and 9.9 are now shown to be consistent in that the computations done in the lab frame and the corresponding ones done in the rocket frame are themselves related by the Lorentz transformation. To do this we proceed as follows.

In the lab the computations are given by

$$x_{k+1} = x_k + (\Delta t_k) v_k \tag{9.50}$$

$$v_{k+1} = \frac{v_k + (\Delta t_k)(1 - v_k^2) F_k \sqrt{1 - v_k^2 + (\Delta t_k)^2 (1 - v_k^2)^2 F_k^2}}{1 + (\Delta t_k)^2 (1 - v_k^2)^2 F_k^2} \tag{9.51}$$

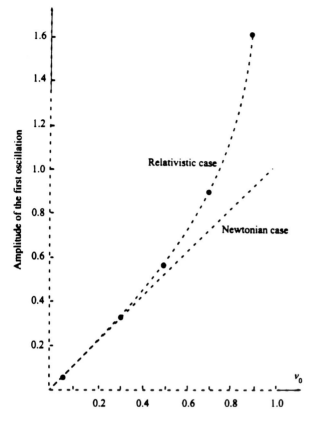

Fig. 9.6 Comparision of Newtonian and relativistic amplitudes.

in which

$$\Delta t_k = t_{k+1} - t_k$$
$$F_k = f(x_k)$$

By the axiom of continuity, calculation in the rocket frame would use

$$x'_{k+1} = x'_k + (\Delta t'_k) v'_k \tag{9.52}$$

$$v'_{k+1} = \frac{v'_k + (\Delta t'_k)(1 - (v'_k)^2) F'_k \sqrt{1 - (v'_k)^2 + (\Delta t'_k)^2 (1 - (v'_k)^2)^2 (F'_k)^2}}{1 + (\Delta t'_k)^2 (1 - (v'_k)^2)^2 (F'_k)^2} \tag{9.53}$$

in which

$$\Delta t'_k = t'_{k+1} - t'_k$$
$$F'_k = f\left(\frac{x'_k + u t'_k}{(1 - u^2)^{\frac{1}{2}}}\right) = f(x_k) = F_k.$$

We want to prove that

$$x'_{k+1} = \frac{(x_{k+1} - ut_{k+1})}{(1-u^2)^{\frac{1}{2}}} \qquad (9.54)$$

$$v'_{k+1} = \frac{v_{k+1} - u}{1 - uv_{k+1}}, \qquad (9.55)$$

which will establish that the computational results in the lab and the rocket are related by the Lorentz transformation.

We assume that x_0, x'_0, v_0, v'_0, the given initial data, are related by the Lorentz transformation. We will then prove (9.54) and (9.55) for $k = 0$. This will be sufficient, by induction, to establish (9.54) and (9.55) in general. Thus

$$x'_0 = \frac{(x_0 - ut_0)}{(1-u^2)^{\frac{1}{2}}}$$

$$v'_0 = \frac{v_0 - u}{1 - uv_0}.$$

Now,

$$x'_1 = x'_0 + (\Delta t'_0)v'_0. \qquad (9.56)$$

Let x^* correspond to x'_1 by the Lorentz transformation, so that

$$x'_1 = \frac{x^* - ut_1}{(1-u^2)^{\frac{1}{2}}}$$

We want to show that $x^* = x_1$.
We know in addition that

$$t'_0 = \frac{t_0 - ux_0}{(1-u^2)^{\frac{1}{2}}}, \quad t'_1 = \frac{t_1 - ux^*}{(1-u^2)^{\frac{1}{2}}}.$$

Thus,

$$\frac{x^* - ut_1}{(1-u^2)^{\frac{1}{2}}} = x'_0 + (\Delta t'_0)v'_0$$

$$= \frac{x_0 - ut_0}{(1-u^2)^{\frac{1}{2}}} + \left(\frac{v_0 - u}{1 - uv_0}\right)\left(\frac{t_1 - ux^*}{(1-u^2)^{\frac{1}{2}}} - \frac{t_0 - ux_0}{(1-u^2)^{\frac{1}{2}}}\right).$$

Since $(1-u^2) > 0$, it then follows that

$$x^* - ut_1 = x_0 - ut_0 + \left(\frac{v_0 - u}{1 - uv_0}\right)(t_1 - t_0 + ux_0 - ux^*),$$

or,

$$(x^* - ut_1)(1 - uv_0)$$
$$= (x_0 - ut_0)(1 - uv_0) + (v_0 - u)(t_1 - t_0) + (v_0 - u)(ux_0 - ux^*),$$

which simplifies to
$$(x^* - x_0)(1 - u^2) = v_0(1 - u^2)(t_1 - t_0).$$
Then,
$$(x^* - x_0) = v_0(t_1 - t_0)$$
so that
$$x^* = x_0 + v_0(t_1 - t_0).$$
Thus, $x^* = x_1$.

Now, for $k = 0$,
$$v_1' = \frac{v_0' + (\Delta t_0')(1 - (v_0')^2)F_0'\sqrt{1 - (v_0')^2 + (\Delta t_0')^2(1 - (v_0')^2)^2(F_0')^2}}{1 + (\Delta t_0')^2(1 - (v_0')^2)^2(F_0')^2}. \quad (9.57)$$

Substitution of
$$t_0' = \frac{t_0 - ux_0}{(1 - u^2)^{\frac{1}{2}}}, \quad t_1' = \frac{t_1 - ux_1}{(1 - u^2)^{\frac{1}{2}}}$$
$$F_0' = F_0 = \frac{v_1 - v_0}{(t_1 - t_0)(1 - v_0^2)(1 - v_1^2)^{\frac{1}{2}}}$$
$$x_0' = \frac{x_0 - ut_0}{(1 - u^2)^{\frac{1}{2}}}, \quad v_0' = \frac{v_0 - u}{1 - uv_0}$$

into (9.57) yields
$$v_1' = \frac{(v_1 - u)\left[(1 - v_0 v_1)(1 - u v_0) + (v_1 - v_0)(u - v_0)\right]}{(1 - uv_1)\left[(1 - v_0 v_1)(1 - u v_0) + (v_1 - v_0)(u - v_0)\right]}.$$

Finally, we will have the desired result
$$v_1' = \frac{(v_1 - u)}{(1 - uv_1)}$$
provided the terms in the brackets, which are identical, are not zero, and we will show this next.

Note that since $|u| < 1$, $|v| < 1$, then
$$|u - v| < 1 - uv.$$
This follows since
$$(1 - u^2)(1 - v^2) > 0$$
$$(u^2 - 1)(1 - v^2) < 0$$
$$u^2 + v^2 - 1 - u^2 v^2 < 0$$
$$u^2 + v^2 < 1 + u^2 v^2$$
$$u^2 - 2uv + v^2 < 1 + u^2 - 2uv$$
$$(u - v)^2 < (1 - uv)^2$$

and
$$|u - v| < |1 - uv| = 1 - uv.$$
However,
$$(1 - v_0 v_1)(1 - uv_0) + (v_1 - v_0)(u - v_0) > 0,$$
since
$$|v_1 - v_0| < 1 - v_0 v_1$$
$$|u - v_0| < 1 - uv_0,$$
so that
$$-(v_1 - v_0)(u - v_0) \leq |v_1 - v_0| \cdot |u - v_0| < (1 - v_0 v_1)(1 - uv_0).$$
Hence
$$(1 - v_0 v_1)(1 - uv_0) + (v_1 - v_0)(u - v_0) > 0.$$

9.11
Remarks

Let us merely indicate here two of the very important results from relativity theory which are available nowhere else. The first is a nondynamical result and was not pertinent in the dynamical studies in previous sections. That is, if one introduces the concept of rest energy, denoted by E, then special relativity establishes that any object, say an atom, which is not in motion, has an inherently large energy of $E = mc^2$, in which m is the mass of the atom. This energy is intimately associated with atomic energy. No such result is implied by Newtonian mechanics or quantum mechanics.

The second result is a dynamical result which would require far too extensive a discussion in order to be proved. It involves what is known as the advance of the perihelion of the planet Mercury. Observations of this planet's motion over a very extensive time period showed that Mercury's orbit around the sun was very close to an ellipse, but was not exactly an ellipse. Indeed, it was established that over a period of 100 years the major axis of Mercury actually rotated 43 seconds, which is called the advance of the perihelion of Mercury. This result was an enigma to astronomers. Now, in Newtonian theory the motion of the planet Mercury around a fixed sun has the equation

$$\frac{d^2 u}{d\theta^2} + u = \frac{\mu}{h^2}$$

in which the various constants are those usually associated with central orbits (Danby (1962) pp. 62–66). Note, however, that since the planet can be considered a single body, and not part of a larger system, the restrictions discussed in Section 9.1, are not applicable. Thus Einstein applied general relativity and merely extended the above equation to

$$\frac{d^2u}{d\theta^2} + u = \frac{\mu}{h^2} + 3\frac{\mu}{c^2}u^2,$$

the solution of which contained the desired 43 seconds in the perihelion motion.

9.12 Exercises

9.1 Show that (9.1a) and (9.1b) are equivalent.

9.2 Provide the details of the proof of Theorem 9.2.

9.3 Prove that (9.44) is the unique solution of (9.40).

9.4 Generate Figure 9.4.

9.5 Generate Figure 9.5.

9.6 Generate Figure 9.6.

10
Special Topics

10.1
Introduction

In this chapter we describe some popular numerical methods which have not been discussed and may be of value in special cases.

10.2
Solving Boundary Value Problems by Initial Value Techniques

Another popular method for solving boundary value problems is called the *shooting method*. In it one utilizes an initial value method in the following way. One guesses $y'(a) = \gamma_1$ and solves the initial value problem

$$y'' = f(x, y, y') \tag{10.1}$$
$$y(a) = \alpha, \quad y'(a) = \gamma_1. \tag{10.2}$$

As shown in Figure 10.1, one computes until $x = b$ and compares the numerical result with the desired result $y = \beta$. The figure also suggests that, physically, one has shot a projectile from the point (a, α) and its path at $x = b$ is below the desired height β. So, one next adjusts the initial angle γ_1 to, say, γ_2, in which $\gamma_2 > \gamma_1$. The numerical calculations are repeated with γ_1 replaced by γ_2 in (10.2). If the numerical solution this time at $x = b$ is, say, greater than β at $x = b$, then one repeats the process with a new angle γ_3 in the range $\gamma_1 < \gamma_3 < \gamma_2$. Hopefully, proceeding with such a process of refinement leads to numerical results which converge to $y = \beta$ at $x = b$. One method of implementation is described in the following example.

Example 10.1 Consider the boundary value problem

$$y'' - y = 0, \quad y(0) = 0, \quad y(1) = 1. \tag{10.3}$$

The shooting method described above is implemented in the following fashion. Kutta's formulas are employed with $h = 0.1$ and the numerical solution

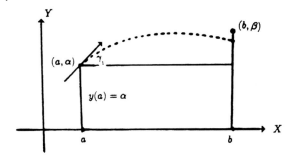

Fig. 10.1 A shooting method.

is generated with each of $\gamma_1 = 0.0, 0.1, 0.2, 0.3, 0.4, 0.5, 0.6, 0.7, 0.8, 0.9, 1.0$. The numerical solution for $\gamma_1 = 0.8$ is less than unity while that for 0.9 is greater than unity at $x = 1$. The Kutta formulas are then reapplied with $\gamma_2 = 0.80, 0.81, 0.82, 0.83, 0.84, 0.85, 0.86, 0.87, 0.88, 0.89, 0.90$. For $\gamma_2 = 0.85$ the numerical solution is less than unity while for $\gamma_2 = 0.86$ it is greater than unity at $x = 1$. Continuing in this fashion yields the approximation $y = 0.999\,99$ for $\gamma_6 = 0.850\,917$ at $x = 1$ while $\gamma_6 = 0.850\,918$ yields a value slightly larger than unity. At this point the numerical solution with $\gamma_6 = 0.850\,917$ is take to be the numerical solution.

Shooting methods are not completely reliable for nonlinear problems and a large number of variations exist (Fertziger (1981); Roberts and Shipman (1972)).

10.3
Solving Initial Value Problems by Boundary Value Techniques

At times it can be advantageous to solve initial value problems by boundary value techniques. This may be the case, for example, when numerical instability for an initial value technique is a problem, or when a problem is stiff. In order to be able to introduce boundary value methods, one needs to be able to determine a second value of y for some value of x, and this is often possible by means of an asymptotic estimate.

Example 10.2 Consider the following initial value problem for the Airy equation:

$$y'' - xy = 0 \qquad (10.4)$$
$$y(0) = 0.355\,028\,053\,887\,817 \qquad (10.5)$$
$$y'(0) = -0.258\,819\,403\,792\,807. \qquad (10.6)$$

For this problem, it is known (Abramowitz and Stegun (1965)) that

$$\left.\begin{array}{l} y(x) > 0, \quad x > 0 \\ \lim_{x \to \infty} y(x) = 0. \end{array}\right\} \tag{10.7}$$

The initial value problem (10.4)–(10.6) is the then replaced by the boundary value problem

$$y'' - xy = 0$$
$$y(0) = 0.355\,028\,053\,887\,817$$
$$y(\infty) = 0,$$

which, in turn, is approximated by the boundary value problem

$$y'' - xy = 0 \tag{10.8}$$
$$y(0) = 0.355\,0280\,538\,878\,17 \tag{10.9}$$
$$y(10\pi) = 0. \tag{10.10}$$

This last boundary value problem is then solved by the method of Section 8.3 with $h = 10\pi/200$. The numerical solution agrees to at least three decimal places with the tabular entries of Abramowitz and Stegun (1965).

10.4
Predictor-Corrector Methods

Another popular class of methods for initial value problems are predictor-corrector methods. The basic idea is as follows. For the initial value problem

$$y' = f(x, y), \quad y(0) = \alpha \tag{10.11}$$

consider the simple Runge–Kutta formula given by (2.10), that is,

$$y_{i+1} = y_i + \frac{1}{2}hf(x_i, y_i) + \frac{1}{2}hf(x_{i+1}, y_i + hf(x_i, y_i)). \tag{10.12}$$

Suppose one rewrites (10.12) as

$$\bar{y}_{i+1} = y_i + hf(x_i, y_i) \tag{10.13}$$

$$y_{i+1} = y_i + \frac{1}{2}h\left[f(x_i, y_i) + f(x_{i+1}, \bar{y}_{i+1})\right]. \tag{10.14}$$

Equation (10.13) can be interpreted as a predictor for y_{i+1} which is determined by Euler's formula, while (10.14) can be interpreted as a corrector for the result $\bar{y}_{i+1}$. If the formulas (10.13), (10.14) are rewritten then as

$$y_{i+1}^{(k)} = y_i + hf(x_i, y_i) \tag{10.15}$$

$$y_{i+1}^{(k+1)} = y_i + \frac{1}{2}h\left[f(x_i, y_i) + f(x_{i+1}, y_{i+1}^{(k)})\right], \quad k = 1, 2, \ldots, \tag{10.16}$$

then the iteration defined by the predictor-corrector formulas (10.15), (10.16) would continue until $y_{i+1}^{(k+1)} = y_{i+1}^{(k)}$, for some fixed value of k, at which time one sets $y_{i+1} = y_{i+1}^{(k)}$.

One of the more popular predictor-corrector formulas, which is comparable to other fourth-order formulas, is the Adams-Moulten pair (Young and Gregory (1973)):

$$y_{i+1}^{(P)} = y_i + \frac{h}{24}\Big[55f(x_i, y_i) - 55f(x_{i-1}, y_{i-1}) \\ + 37f(x_{i-2}, y_{i-2}) - 9f(x_{i-3}, y_{i-3})\Big] \quad (10.17)$$

$$y_{i+1} = y_i + \frac{h}{24}\Big[9f(x_{i+1}, y_{i+1}^{(P)}) + 19f(x_i, y_i) \\ - 5f(x_{i-1}, y_{i-1}) + f(x_{i-2}, y_{i-2})\Big]. \quad (10.18)$$

It should be noted that the use of (10.17), (10.18) requires knowing $y_{i-1}, y_{i-2}, y_{i-3}$. Thus, some other method is required to determine y_1 and y_2 before the Adams-Moulten method can be implemented.

10.5
Multistep Methods

The methods of Chapters 2 and 3 for initial value problems are called *one step* methods, because each generates y_{i+1} using no other previously calculated values of y except y_i. Methods which use additional previously calculated values of y are called multistep methods. A typical multistep formula (Jain (1984)) for

$$y' = f(x, y), \quad y(0) = \alpha$$

is, for $i = 2, 3, 4, \ldots,$

$$y_{i+1} = y_i + \frac{1}{12}h(23f(x_i, y_i) - 16f(x_{i-1}, y_{i-1}) + 5f(x_{i-2}, y_{i-2})). \quad (10.19)$$

Note that multistep formula (10.19) cannot be implemented until y_1 and y_2 have been generated by other methods.

A great number of results concerning the stability of linear multistep methods are available (Dahlquist and Bjork (1974)).

10.6
Other Methods

A large number of additional methods are available for the numerical solution of ordinary differential equations problems. These include the method of isoclines; backward Euler; extrapolation; high order interpolation; trigonometric, Chebyshev and Lie series; Monte Carlo simulation; and finite elements. Related references are contained in the References.

10.7
Consistency*

Consistency is simply the condition that if one takes the limit as h goes to zero of a difference equation, then that limit converges to the differential equation it is intended to approximate. For example, consider the differential equation

$$y'' + 3y' - 2y = x \tag{10.20}$$

and the difference equation approximation

$$\frac{y_{i-1} - 2y_i + y_{i+1}}{h^2} + 3\frac{y_{i+1} - y_{i-1}}{2h} - 2y_i = x_i. \tag{10.21}$$

As h converges to zero in (10.21), the difference equation converges to the differential equation in (10.20) and is said to be a consistent approximation of (10.20).

It is most interesting to observe that one can have a consistent approximation to a differential equation and yet, as h converges to zero, the numerical solution will not converge to the solution of the differential equation. For example, consider the initial value problem

$$y' = -100y, \quad y(0) = 1. \tag{10.22}$$

For any h, consider the grid points $x_0, x_1, x_2, \ldots$. Then, $y_0 = 1$. Approximate y_1 by Euler's method so that $y_1 = 1 - 100h$. Thereafter, approximate $y_2, y_3, y_4, \ldots$ by

$$\frac{y_{i+2} - y_i}{2h} = -100y_{i+1}. \tag{10.23}$$

As h converges to zero, (10.23) converges to (10.22) and is, therefore, a consistent approximation.

Now, (10.23) can be rewritten as the two-step formula

$$y_{i+2} = y_i - 200hy_{i+1}, \quad i = 0, 1, 2, 3, \ldots$$

which is a linear, second-order difference equation. To determine the stability of the equation, one rewrites it as

$$y_{i+2} + 200hy_{i+1} - y_i = 0 \tag{10.24}$$

and one sets $y_i = \lambda^i$, $\lambda \neq 0$. Thus

$$\lambda^2 + 200h\lambda - 1 = 0. \tag{10.25}$$

The two roots of (10.25) are

$$\lambda_1 = -100h + \sqrt{1 + (100h)^2}, \quad \lambda_2 = -100h - \sqrt{1 + (100h)^2}.$$

The general solution of (10.24) is then

$$y_i = c_1(\lambda_1)^i + c_2(\lambda_2)^i, \quad i = 2, 3, 4, \ldots, \tag{10.26}$$

in which c_1, c_2 are arbitrary constants. Now, c_1 and c_2 can be determined from y_0 and y_1. Indeed, $c_2 = \frac{-1 + \sqrt{1 + (100h)^2}}{2\sqrt{1 + (100h)^2}} \neq 0$, while $|\lambda_2| > 1$. Thus, for all h, (10.26) is unstable.

However, the solution of (10.22) is $y = e^{-100x}$, so that the exact solution converges to zero as x goes to infinity. Thus, convergence is not possible for any choice of h.

For linear multistep methods, a more general definition of consistency, which includes the one above, and a definition of stability, which is different from the one we have adopted, yield the following interesting theorem (Dahlquist and Bjork (1974)):

Theorem 10.1 *Consistency and stability imply convergence, and convergence implies consistency and stability.*

10.8
Differential Eigenvalue Problems

In studies of atoms, molecules, bending beams, and structural stability, the following type of problem is of interest and is called a differential eigenvalue problem.

For what positive values of λ does the boundary value problem

$$y'' + \lambda^2 y = 0, \quad y(0) = y(\pi) = 0, \quad (\lambda > 0) \tag{10.27}$$

have nonzero solutions? Such a problem is called a differential eigenvalue problem. The nonzero, positive values λ are called eigenvalues and the associated solutions of the differential equation are called eigenvectors. Note, in particular, that (10.27) need not have a unique solution because condition (8.14) is not satisfied.

Because of its simplicity, (10.27) can be solved exactly as follows. The general solution of the differential equation is

$$y(x) = c_1 \sin(\lambda x) + c_2 \cos(\lambda x). \tag{10.28}$$

However, the boundary conditions imply only that $c_2 = 0$, so that

$$y(x) = c_1 \sin(\lambda x). \tag{10.29}$$

The boundary conditions imply also, with respect to (10.27), that λ can take on any of the values $\lambda = 1, 2, 3, \ldots$, and only these values, since $(\sin \lambda x)$ must be zero when $x = \pi$. Thus, the eigenvalues are $\lambda = 1, 2, 3, \ldots$, and the eigenvectors are given by (10.29), where c_1 is any nonzero constant.

If the differential equation in (10.27) was, however, more complex, then one might not be able to carry through the exact analysis, as was done above. In such cases, one could approximate the differential eigenvalue problem solutions by adapting one of the numerical methods already described, the choice being dependent on the physical results desired. To illustrate, suppose one were interested primarily in the minimum eigenvalue of (10.27), which is often of interest in atomic and molecular studies. One could apply a difference method as follows.

Divide the interval $0 \leq x \leq \pi$ into four equal parts by the points $x_0 = 0$, $x_1 = \pi/4$, $x_2 = \pi/2$, $x_3 = 3\pi/4$, $x_4 = \pi$. At each interior grid point approximate (10.27) by

$$y_{i-1} - 2y_i + y_{i+1} + (\lambda \pi/4)^2 y_i = 0.$$

Setting $\mu = (\lambda \pi/4)^2$ and $i = 1, 2, 3$, yields the three equations

$$\begin{aligned}
(\mu - 2)y_1 + y_2 &= 0 \\
y_1 + (\mu - 2)y_2 + y_3 &= 0 \\
y_2 + (\mu - 2)y_3 &= 0
\end{aligned}$$

For this system to have a nonzero solution, the determinant of the system must be zero, or, equivalently,

$$(\mu - 2)^3 - 2(\mu - 2) = 0$$

which yields

$$\mu = 2 - 2^{\frac{1}{2}}, 2, 2 + 2^{\frac{1}{2}}.$$

The minimum root μ is then $2 - 2^{\frac{1}{2}}$, which yields, approximately, a minimum λ given by

$$\lambda = 0.974. \tag{10.30}$$

The exact solution, as described above, is $\lambda = 1$.

In order to improve on (10.30), one need only decrease the grid size. If, for example, one halves the grid size so that $h = \pi/8$, and then repeats the procedure with $\mu = (\lambda \pi/8)^2$, then the equation one has to solve for μ is

$$\mu^7 - 14\mu^6 + 78\mu^5 - 220\mu^4 + 330\mu^3 - 252\mu^2 + 84\mu - 8 = 0. \tag{10.31}$$

Computer evaluation of the left side of (12.31) for $\mu = 0, 0.1, 0.2, \ldots, 9.9, 10$ reveals quickly by the sign changes that all seven roots of (12.31) are real, positive, and lie in the interval $0.1 \leq \mu \leq 3.9$. In particular, the smallest root lies in $0.1 \leq \mu \leq 0.2$. Application of Newton's method on this interval yields $\mu = 0.152$, from which it follows that the minimum λ is

$$\lambda = 0.993,$$

which is a significant improvement over (10.30). Note that use of the grid size $\pi/4$ yields only three approximate eigenvalues, while use of the grid size $\pi/8$ yields seven approximations. In general, the division of $0 \leq x \leq \pi$ into n parts results in $n - 1$ approximate eigenvalues, even though the number of exact eigenvalues is infinite.

10.9
Chaos*

At present, extensive interest and energy are directed to a phenomenon called chaos. Our purpose in this section is merely to place these activities, which center around systems of ordinary differential equations, in perspective.

The most prevalent and enigmatic form of fluid behavior is turbulence (von Karman (1962)). Turbulence results when the molecules of a fluid no longer flow in a laminar fashion. The fundamental equations of fluid dynamics, the Navier–Stokes equations, are equations of laminar flow (Schlichting (1960)). These equations are highly nonlinear partial differential equations, and since they are laminar, cannot be the equations of turbulence. Nevertheless, there are some researchers who assume that these equations represent turbulence in some average sense. Under this assumption, the term chaos is often used interchangeably with turbulence (Barenblatt et al. (1983)). To simulate chaos, Lorenz (1963), in studying a particular set of Navier–Stokes equations, tried to solve them using a Galerkin expansion. He truncated the expansion drastically enough to deduce three ordinary differential equations with bilinear nonlinearities. Originally, all the results on strange attractors and chaotic behavior were derived in analyzing these three equations. Though related computer simulations have yielded interesting flow patterns, it is not clear at present to what extent these simulations correspond to actual turbulent behavior.

10.10
Contact Mechanics

One particular area of mechanics which has received extensive study in physics and engineering is contact mechanics. A particular example of such a study has already been explored on the molecular level in Section 4.17. An example in the large of a bouncing elastic ball is shown in Figure 10.2

(Greenspan (2004)). A very large number of particular models and studies are available in the literature (see, e.g., Shillor (1988); Lebon and Raous (1992)). However, the problems of contact mechanics, though most important, are so specific and of such difficulty that as yet the field awaits a comprehensive underlying theory.

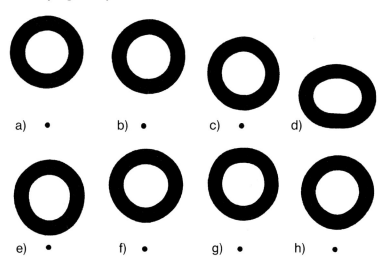

Fig. 10.2 A bouncing elastic ball.

Appendix A
Basic Matrix Operations

For a positive number n, an n-dimensional column vector is simply a vertical array of n numbers written

$$\begin{bmatrix} a_1 \\ a_2 \\ a_3 \\ \vdots \\ a_n \end{bmatrix}.$$

An example of a three-dimensional column vector is

$$\begin{bmatrix} -2 \\ 7 \\ 0 \end{bmatrix}.$$

Two n-dimensional column vectors are said to be equal if and only if they are termwise equal. Thus if

$$\begin{bmatrix} 1 \\ 3 \\ -2 \end{bmatrix} = \begin{bmatrix} a \\ x \\ \beta \end{bmatrix},$$

then $a = 1$, $x = 3$, $\beta = -2$.

For positive integer n, an n-dimensional matrix is simply an $n \times n$, array of numbers written

$$\begin{bmatrix} a_{11} & a_{12} & a_{13} & \cdots & a_{1,n} \\ a_{21} & a_{22} & a_{23} & \cdots & a_{2,n} \\ a_{31} & a_{32} & a_{33} & \cdots & a_{3n} \\ \vdots & & & & \\ a_{n1} & a_{n2} & a_{n3} & \cdots & a_{nn} \end{bmatrix}.$$

Numerical Solution of Ordinary Differential Equations for Classical, Relativistic and Nano Systems. Donald Greenspan
Copyright © 2006 WILEY-VCH Verlag GmbH & Co. KGaA, Weinheim
ISBN: 3-527-40610-7

An example of a three-dimensional matrix is

$$\begin{bmatrix} 2 & 0 & -6 \\ -4 & 1 & 3.1 \\ 7 & 9 & 2 \end{bmatrix}.$$

Two n-dimensional matrices are said to be equal if and only if they are equal termwise. Thus,

$$\begin{bmatrix} 2 & 0 & -6 \\ -4 & 1 & 3.1 \\ 7 & 9 & 2 \end{bmatrix} = \begin{bmatrix} a_{11} & a_{12} & a_{13} \\ a_{21} & a_{22} & a_{23} \\ a_{31} & a_{32} & a_{33} \end{bmatrix},$$

then $a_{11} = 2$, $a_{12} = 0$, $a_{13} = -6$, $a_{21} = -4$, $a_{22} = 1$, $a_{23} = 3.1$, $a_{31} = 7$, $a_{32} = 9$, $a_{33} = 2$.

The product

$$\begin{bmatrix} a_{11} & a_{12} & a_{13} & \cdots & a_{1,n} \\ a_{21} & a_{22} & a_{23} & \cdots & a_{2,n} \\ a_{31} & a_{32} & a_{33} & \cdots & a_{3n} \\ \vdots & & & & \\ a_{n1} & a_{n2} & a_{n3} & \cdots & a_{nn} \end{bmatrix} \begin{bmatrix} a_1 \\ a_2 \\ a_3 \\ \vdots \\ a_n \end{bmatrix}$$

is defined to be the vector

$$\begin{bmatrix} a_{11}a_1 + a_{12}a_2 + \cdots + a_{1n}a_n \\ a_{21}a_1 + a_{22}a_2 + \cdots + a_{2n}a_n \\ a_{31}a_1 + a_{32}a_2 + \cdots + a_{3n}a_n \\ \vdots \\ a_{n1}a_1 + a_{n2}a_2 + \cdots + a_{nn}a_n \end{bmatrix}.$$

Thus,

$$\begin{bmatrix} 3 & 1 & -1 \\ 2 & 0 & 1 \\ -1 & 3 & 2 \end{bmatrix} \begin{bmatrix} 1 \\ 2 \\ 3 \end{bmatrix} = \begin{bmatrix} 3 \cdot 1 + 1 \cdot 2 - 1 \cdot 3 \\ 2 \cdot 1 + 0 \cdot 2 + 1 \cdot 3 \\ -1 \cdot 1 + 3 \cdot 2 + 2 \cdot 3 \end{bmatrix} = \begin{bmatrix} 2 \\ 5 \\ 11 \end{bmatrix}.$$

The sum of two n-dimensional matrices is defined to be their termwise sum. Thus

$$\begin{bmatrix} 3 & 2 & -7 \\ 0 & 1 & -4 \\ 6 & 0 & 1 \end{bmatrix} + \begin{bmatrix} -1 & 1 & 0 \\ 0 & 2 & 0 \\ 1 & 0 & 1 \end{bmatrix} = \begin{bmatrix} 2 & 3 & -7 \\ 0 & 3 & -4 \\ 7 & 0 & 2 \end{bmatrix}.$$

The product of two n-dimensional matrices

$$\begin{bmatrix} a_{11} & a_{12} & a_{13} & \cdots & a_{1,n} \\ a_{21} & a_{22} & a_{23} & \cdots & a_{2,n} \\ a_{31} & a_{32} & a_{33} & \cdots & a_{3n} \\ \vdots & & & & \\ a_{n1} & a_{n2} & a_{n3} & \cdots & a_{nn} \end{bmatrix} \cdot \begin{bmatrix} b_{11} & b_{12} & b_{13} & \cdots & b_{1,n} \\ b_{21} & b_{22} & b_{23} & \cdots & b_{2,n} \\ b_{31} & b_{32} & b_{33} & \cdots & b_{3n} \\ \vdots & & & & \\ b_{n1} & b_{n2} & b_{n3} & \cdots & b_{nn} \end{bmatrix}$$

is defined to be the n-dimensional matrix whose jth column is

$$\begin{bmatrix} a_{11} & a_{12} & a_{13} & \cdots & a_{1,n} \\ a_{21} & a_{22} & a_{23} & \cdots & a_{2,n} \\ a_{31} & a_{32} & a_{33} & \cdots & a_{3n} \\ \vdots & & & & \vdots \\ a_{n1} & a_{n2} & a_{n3} & \cdots & a_{nn} \end{bmatrix} \begin{bmatrix} b_{1j} \\ b_{2j} \\ b_{3j} \\ \vdots \\ b_{nj} \end{bmatrix}.$$

Thus,

$$\begin{bmatrix} 3 & 0 & -1 \\ 1 & 2 & 0 \\ 2 & 0 & 3 \end{bmatrix} \cdot \begin{bmatrix} -1 & 1 & 0 \\ 0 & 1 & 0 \\ 1 & 0 & 2 \end{bmatrix} = \begin{bmatrix} -4 & 3 & -2 \\ -1 & 3 & 0 \\ 1 & 2 & 6 \end{bmatrix}.$$

Note that throughout, we assume the numbers in vectors and matrices, that is, the elements, are real numbers.

Solutions to Selected Exercises

Chapter 1

1.3 $y_{exact} = 4/(3e^{2x} + 2x + 1)$.

x	y_{num}	y_{exact}	$|E|$
0.05	0.900000	0.905897	0.005897
0.15	0.734119	0.747723	0.013603
0.25	0.602693	0.620524	0.017831
0.35	0.496756	0.516712	0.019956
0.45	0.410334	0.431090	0.020756
0.55	0.339260	0.359955	0.020695
0.65	0.280504	0.300574	0.020069
0.75	0.213786	0.250861	0.019075
0.85	0.191334	0.209185	0.017850
0.95	0.157740	0.174234	0.016494

1.7 (a) $M = 1$, $N = 3$.

Chapter 2

2.4 Table entries rounded from 17 decimal places.

x	y_{sar}	y_{exact}	$|E|$
0.20	0.81873075306	0.81873075307	0.00000000001
0.40	0.67032004601	0.67032004603	0.00000000002
0.60	0.54881163606	0.54881163609	0.00000000003
0.80	0.44932896408	0.44932896412	0.00000000004
1.00	0.36787944113	0.36787944117	0.00000000004
1.20	0.30119421187	0.30119421191	0.00000000004

Numerical Solution of Ordinary Differential Equations for Classical, Relativistic and Nano Systems. Donald Greenspan
Copyright © 2006 WILEY-VCH Verlag GmbH & Co. KGaA, Weinheim
ISBN: 3-527-40610-7

2.6 Second and third column entries rounded from 18 decimal places.

x	y_{shanks}	y_{exact}	$|E|$
0.40	0.67032004603	0.67032004603	0.000000000000000098
0.80	0.44932896411	0.44932896411	0.000000000000000100
1.20	0.30119421191	0.30119421191	0.000000000000000087
1.60	0.20189651799	0.20189651799	0.000000000000000068
2.00	0.13533528323	0.13533528323	0.000000000000000049

2.12 (a) and (b)

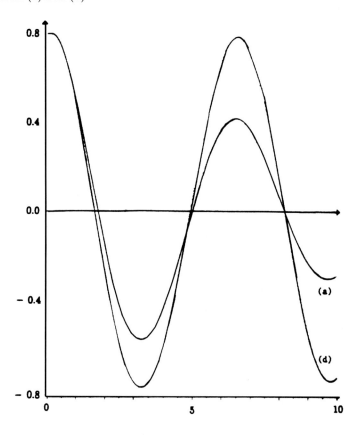

Chapter 3

3.2 Second and third columns rounded from thirteen decimal places.

x	y_{Taylor}	y_{exact}	$|E|$
0.40	0.670320	0.670320	0.0000000001613
0.80	0.449329	0.449329	0.0000000002664
1.20	0.301194	0.301194	0.0000000003019
1.60	0.201897	0.201897	0.0000000002865
2.00	0.135335	0.135335	0.0000000002453

Chapter 5

5.1

$$m_1\ddot{x}_1 = -\frac{Gm_1m_2}{r_{12}^2}\frac{x_1-x_2}{r_{12}} - \frac{Gm_1m_3}{r_{13}^2}\frac{x_1-x_3}{r_{13}}$$

$$m_1\ddot{y}_1 = -\frac{Gm_1m_2}{r_{12}^2}\frac{y_1-y_2}{r_{12}} - \frac{Gm_1m_3}{r_{13}^2}\frac{y_1-y_3}{r_{13}}$$

$$m_2\ddot{x}_2 = -\frac{Gm_1m_2}{r_{12}^2}\frac{x_2-x_1}{r_{12}} - \frac{Gm_2m_3}{r_{23}^2}\frac{x_2-x_3}{r_{23}}$$

$$m_2\ddot{y}_2 = -\frac{Gm_1m_2}{r_{12}^2}\frac{y_2-y_1}{r_{12}} - \frac{Gm_2m_3}{r_{23}^2}\frac{y_2-y_3}{r_{23}}$$

$$m_3\ddot{x}_3 = -\frac{Gm_1m_3}{r_{13}^2}\frac{x_3-x_1}{r_{13}} - \frac{Gm_2m_3}{r_{23}^2}\frac{x_3-x_2}{r_{23}}$$

$$m_3\ddot{y}_3 = -\frac{Gm_1m_3}{r_{13}^2}\frac{y_3-y_1}{r_{13}} - \frac{Gm_2m_3}{r_{23}^2}\frac{y_3-y_2}{r_{23}},$$

in which

$$r_{ij}^2 = (x_i - x_j)^2 + (y_i - y_j)^2.$$

$$x_{i,n+1} - x_{i,n} - \frac{1}{2}(\Delta t)(v_{i,n+1,x} + v_{i,n,x}) = 0$$

$$y_{i,n+1} - y_{i,n} - \frac{1}{2}(\Delta t)(v_{i,n+1,y} + v_{i,n,y}) = 0$$

$$v_{i,n+1,x} - v_{i,n,x} - \frac{\Delta t}{m_i}F_{i,n,x} = 0$$

$$v_{i,n+1,y} - v_{i,n,y} - \frac{\Delta t}{m_i}F_{i,n,y} = 0,$$

in which the F's are given by

$$F_{1,n,x} = -\frac{Gm_1m_2[(x_{1,n+1}+x_{1,n})-(x_{2,n+1}+x_{2,n})]}{r_{12,n}r_{12,n+1}[r_{12,n}+r_{12,n+1}]}$$
$$-\frac{Gm_1m_3[(x_{1,n+1}+x_{1,n})-(x_{3,n+1}+x_{3,n})]}{r_{13,n}r_{13,n+1}[r_{13,n}+r_{13,n+1}]}$$

$$F_{1,n,y} = -\frac{Gm_1m_2[(y_{1,n+1}+y_{1,n})-(y_{2,n+1}+y_{2,n})]}{r_{12,n}r_{12,n+1}[r_{12,n}+r_{12,n+1}]}$$
$$-\frac{Gm_1m_3[(y_{1,n+1}+y_{1,n})-(y_{3,n+1}+y_{3,n})]}{r_{13,n}r_{13,n+1}[r_{13,n}+r_{13,n+1}]}$$

$$F_{2,n,x} = -\frac{Gm_1m_2[(x_{2,n+1}+x_{2,n})-(x_{1,n+1}+x_{1,n})]}{r_{12,n}r_{12,n+1}[r_{12,n}+r_{12,n+1}]}$$
$$-\frac{Gm_2m_3[(x_{2,n+1}+x_{2,n})-(x_{3,n+1}+x_{3,n})]}{r_{23,n}r_{23,n+1}[r_{23,n}+r_{23,n+1}]}$$

$$F_{2,n,y} = -\frac{Gm_1m_2[(y_{2,n+1}+y_{2,n})-(y_{1,n+1}+y_{1,n})]}{r_{12,n}r_{12,n+1}[r_{12,n}+r_{12,n+1}]}$$
$$-\frac{Gm_2m_3[(y_{2,n+1}+y_{2,n})-(y_{3,n+1}+y_{3,n})]}{r_{23,n}r_{23,n+1}[r_{23,n}+r_{23,n+1}]}$$

$$F_{3,n,x} = -\frac{Gm_1m_3[(x_{3,n+1}+x_{3,n})-(x_{1,n+1}+x_{1,n})]}{r_{13,n}r_{13,n+1}[r_{13,n}+r_{13,n+1}]}$$
$$-\frac{Gm_2m_3[(x_{3,n+1}+x_{3,n})-(x_{2,n+1}+x_{2,n})]}{r_{23,n}r_{23,n+1}[r_{23,n}+r_{23,n+1}]}$$

$$F_{3,n,y} = -\frac{Gm_1m_3[(y_{3,n+1}+y_{3,n})-(y_{1,n+1}+y_{1,n})]}{r_{13,n}r_{13,n+1}[r_{13,n}+r_{13,n+1}]}$$
$$-\frac{Gm_2m_3[(y_{3,n+1}+y_{3,n})-(y_{2,n+1}+y_{2,n})]}{r_{23,n}r_{23,n+1}[r_{23,n}+r_{23,n+1}]},$$

and

$$r_{ij,m}^2 = (x_{i,m}-x_{j,m})^2 + (y_{i,m}-y_{j,m})^2; \quad m = n, n+1.$$

Chapter 6

6.1 (a) $h \leq 2$
(c) always stable.

6.2 (a) unstable
(c) stable
(e) unstable
(g) stable
(i) stable
(k) unstable
(m) stable.

6.5 Hint: Use $K + \phi = K - \frac{1}{r} = -\epsilon \geq -\frac{1}{r}$.

Chapter 7

7.1 (a) $x_1 = -1$, $x_{100} = -5$, all other x values are 3.

References

M. Abramowitz and I. A. Stegun, Handbook of Mathematical Functions, NBS, Washington, D.C., 1965 (also available on disk).

A. S. Argon and S. Yip, "Molecular dynamics simulation of crack tip processes in alpha-iron and copper", J. Appl. Phys., 54, 1983, p. 48.

A. Aziz (Ed.), Numerical Solutions of Boundary Value Problems in Ordinary Differential Equations, Academic Press, New York, 1975.

G. I. Barenblatt, G. Iooss, and D. D. Joseph, Nonlinear Dynamics and Turbulence, Pitman, Boston, 1983.

R. E. Bellman, R.E. Kalaba, and J. A. Lockett, Numerical Inversion of the Laplace Transform, Elsevier, New York, 1966.

L. Bers, "On mildly nonlinear partial differential equations", J. Res. NBS, 51, 1953, p. 229.

G. Birkhoff and G.-C. Rota, Ordinary Differential Equations, Ginn and Co., Boston, 1962.

W. E. Boyce and R. C. DiPrima, Elementary Differential Equations, 4th Ed., Wiley, New York, 1986.

M. L. Brodskii, "Asymptotic estimates of the errors in the numerical integration of systems of ordinary differential equations by difference methods", Dokl. Akad. SSSR, (N.S.), 93, 1953, p. 599.

R. Burlisch and J. Stoer, "Numerical treatment of ordinary differential equations by extrapolation methods", Num. Mat., 8, 1966, p. 1.

J. C. Butcher, "On Runge–Kutta processes of high order", J. Australian Math. Soc., 4, 1964, p. 179.

J. C. Butcher, "On attainable order of Runge–Kutta methods", Math. Comp., 19, 1965, p. 408.

J. C. Butcher, Numerical Analysis of Ordinary Differential Equations: Runge-Kutta and Linear Methods, Wiley, New York, 1987.

F. Calogero, "Exactly solvable one-dimensional many body problems", Lett. Nuovo Cimento, 11, 1975, p. 411.

V. Capra, "Valutazione degli errori nella integrazione numerica dei systemi di equazioni differenziali ordinarie", Atti Accad. Sci. Torino Cl. Sci. Fis. Mat. Nat., 91, 1957, p. 188.

J. W. Carr, "Error bounds for the Runge–Kutta single step process", J. ACM, 5, 1958, p. 39.

L. Cesari, Asymptotic Behavior and Stability Problems in Ordinary Differential Equations, Springer, Berlin, 1959.

C. W. Clenshaw, "The solution of van der Pol's equation in Chebychev series", in Numerical Solutions of Nonlinear Differential Equations (Ed. D. Greenspan), Wiley, New York, 1966, p. 55.

E. A. Coddington and N. Levinson, Theory of Ordinary Differential Equations, McGraw-Hill, New York, 1955.

L. Collatz, The Numerical Treatment of Differential Equations, Springer, Berlin, 1960.

G. Corliss and Y. F. Chang, "Solving ordinary differential equations using Taylor series", ACM Trans. in Math. Soft., 8, 1982, p. 114.

R. Courant, Differential and Integral Calculus, vols. I and II, Interscience, New York, 1947.

R. Courant and D. Hilbert, Methods of Mathematical Physics, vol. II, Wiley, New York, 1962.

J. Cronin, Fixed Points and Topological Degree in Nonlinear Analysis, AMS, Providence, RI, 1964, Chap. II.

C. W. Cryer, "Stability analysis in discrete mechanics", TR #67, Comp. Sci. Dept., University Wisconsin, Madison, 1969.

A. R. Curtis, "High-order explicit Runge–Kutta formulae, their uses, and limitations", JIMA, 16, 1975, p. 35.

C. F. Curtiss and J. O. Hirschfelder, "Integration of stiff equations", Proc. Natl. Acad. Sci. USA, 38, 1952, p. 235.

G. Dahlquist and A. Bjork, Numerical Methods, Prentice-Hall, Englewood Cliffs, 1974.

J. M. A. Danby, Fundamentals of Celestial Mechanics,Macmillan, New York, 1962.

J. W. Daniel and R. E. Moore, Computation and Theory in Ordinary Differential Equations, Freeman, San Francisco, 1970.

R. de Vogelaere, "A method for the numerical integration of second order differential equations without explicit first derivatives", J. Res. NBS, 54, 1955, p. 119.

J. Dyer, "Generalized multistep methods in satellite orbit computation" J. ACM, 15, 1968, p. 712.

K. H. Ebert, P. Deuflhard, and W. Jager (Eds), Modelling of Chemical Reaction Systems, Springer, Berlin, 1981.

R. England, "Error estimates for Runge–Kutta type solutions to systems of O.D.E.'s", Comp. J., 12, 1969, p. 166.

L. Euler, Opera Omnia, Series Prima, Vol. 11, Leipzig and Berlin, 1913.

L. Euler, Opera Omnia, Series Prima, Vol. 12, Leipzig and Berlin, 1914.

E. Fehlberg, "New high order Runge–Kutta formulas with an arbitrarily small truncation error", ZAMM, 46, 1966, p. 1.

E. Fehlberg, "Classical fifth-, sixth- seventh- and eighth-order Runge–Kutta formulas with step size control", NASA, TR-R-287, Marshall Space Flight Center, Huntsville, Ala., 1968.

R. P. Feynman, R. B. Leighton and M. Sands, The Feynman Lectures on Physics, Addison-Wesley, Reading, 1963.

L. Fox, The Numerical Solution of Two-point Boundary Value Problems in Ordinary Differential Equations, Oxford University Press, New York, 1957.

J. N. Franklin, "Difference methods for stochastic ordinary differential equations", Math. Comp., 19, 1965, p. 552.

C. W. Gear, Numerical Initial Value Problems in Ordinary Differential Equations, Prentice-Hall, Englewood Cliffs, 1971.

G. H. Golub and C. F. van Loan, Matrix Computations, Johns Hopkins University Press, Baltimore, 1983.

D. Greenspan, Theory and Solution of Ordinary Differential Equations, Macmillan, New York, 1960.

D. Greenspan, "Approximate solution of initial value problems by boundary value techniques", J. Math. Phys. Sci., I, 1967, p. 261.

D. Greenspan, Discrete Models, Addison-Wesley, Reading, 1973.

D. Greenspan, Arithmetic Applied Mathematics, Pergamon, Oxford, 1980.

D. Greenspan, Computer-Oriented Mathematical Physics, Pergamon, Oxford, 1981.

D. Greenspan, "Conservative difference formulation of Colagero and Toda Hamiltonian systems", Comp. Math. Applic., 19, 1990, p. 91.

D. Greenspan, Particle Modeling, Birkhauser, Boston, 1997.

D. Greenspan, "Conservative motion of discrete tetrahedral tops and gyroscopes", Appl. Math. Modelling, 22, 1998, p. 57.

D. Greenspan, "Conservative motion of a discrete dodecahedral gyroscope", Math. and Comp. Modelling, 35, 2002, p. 323.

D. Greenspan, N-body Problems and Models, World Scientific Publishing, Singapore, 2004, p. 141.

D. Greenspan, Molecular and Particle Modelling of Laminar and Turbulent Flows, World Scientific Publishing, Singapore (to appear summer 2005).

D. Greenspan and V. Casulli, Numerical Analysis for Applied Mathematics, Science and Engineering, Addison-Wesley, Redwood City, 1988.

D. Greenspan and L. Heath, "Super computer simulation of the modes of colliding micro drops of water", J. Phys. D: Appl. Phys., 24, 1991, p. 2121.

E. Hairer, "A Runge–Kutta method of order 10", J. IMA, 21, 1978, p. 47.

P. Henrici, Discrete Variable Methods in Ordinary Differential Equations, Wiley, New York, 1963.

P. Henrici, Error Propagation in Difference Methods, Wiley, New York, 1963.

K. Heun, "Neue Methode zur approximativen Integration der Differentialgleichungen einer unabhängigen Veränderlichen", ZAMP, 45, 1900, p. 23.

J. O. Hirschfelder, C. F. Curtiss and R. B. Bird, Molecular Theory of Gases and Liquids, Wiley, New York, 1967.

R. W. Hockney and J. W. Eastwood, Computer Simulation Using Particles, McGraw-Hill, New York, 1981.

M. K. Jain, Numerical Solution of Differential Equations, 2nd Ed., Wiley Eastern Ltd., New Delhi, 1984.

E. Kamke, Differentialgleichungen, Akad.-Verlag, Leipzig, 1959.

H. B. Keller, Numerical Methods for Two-point Boundary Value Problems, Blaisdell, Waltham, 1968.

H. Knapp and G. Wanner, "Liese, a program for ordinary differential equations using Lie series", TR #881, MRC, University Wisconsin, Madison, 1968.

A. N. Kolmogorov, "Toward a more precise notion of the structure of the local turbulence in a viscous fluid at elevated Reynolds number", in: THE MECHANICS OF TURBULENCE (Ed. A. Favre), Gordon and Breach, New York, 1964, p. 447.

J. Koplik and J. R. Banavar, "Physics of fluids at low Reynolds number - A molecular approach", Computers and Physics, 12, 1998, p. 424.

A. Korzeniowski and D. Greenspan, "Microscopic turbulence in water", Math. Comput. Modelling, 23, 1996, p. 89.

W. S. Krogdahl, The Astronomical Universe, Macmillan, New York, 1952.

W. Kutta, "Beitrag zur näherungweisen Integration totaler Differentialgleichungen", ZAMP, 46, 1901, p. 435.

R. A. LaBudde and D. Greenspan, "Energy and momentum conserving methods of arbitrary order for the numerical integration of equations of motion. II" Numer. Math., 26, 1976, p. 1.

F. Lebon and M. Raous, "Multibody contact problems including friction in structure assembly", Computers and Structures, 43, 1992, p. 925.

S. Lefshetz, Differential Equations: Geometric Theory, 2nd Ed., Interscience, New York, 1963, Chap. II.

H. Lieberstein, "Over relaxation for nonlinear elliptic partial differential equations", TR #80, MRC, University Wisconsin, Madison, 1959.

E. N. Lorenz, "Deterministic non periodic flow", J. Atmo. Sci., 20, 1963, p. 130.

F. Loscalzo and T. Talbot, "Spline function approximations for solutions of ordinary differential equations", SIAM J. Num. Anal., 4, 1967, p. 433.

A. N. Lowan, "On the propagation of errors in the inversion of certain tridiagonal matrices", Math. Comp., 14, 1960, p. 333.

T. R. Lucus and G. W. Reddien, "Some collocation methods for nonlinear boundary value problems", SIAM J. Num. Anal., 9, 1972, p. 341.

L. Lustman, B. Neta and W. Gragg, "Solution of ordinary differential initial value problems on an INTEL hyper cube", Comp. Math. Appl., 23, 1992, p. 65.

H. A. Luther, "An explicit sixth-order Runge–Kutta formula", Math. Comp., 22, 1968, p. 344.

A. Marciniak, Numerical Solutions of the N-body Problem, Reidel, Dordrecht, 1985.

J. E. Marsden, Lectures on Geometrical Methods in Mathematical Physics, SIAM, Philadelphia, 1981.

W. L. Masterton and E. J. Slowinski, Chemical Principles, 2nd Ed., Saunders, Philadelphia, 1969, p. 96.

J. A. McCammon and S. C. Harvey, Dynamics of Proteins and Nucleic Acids, Cambridge University Press, New York, 1987.

W. Miranker and W. Liniger, "Parallel methods for the numerical integration of ordinary differential equations", Math. Comp., 21, 1967, p. 303.

R. E. Moore, Methods and Applications of Interval Analysis, SIAM, Philadelphia, 1979.

J. Moser, "Various aspects of integrable Hamiltonian systems", CIME Lectures in Dynamical Systems, Birkhauser, Boston, 1980.

A. Nordsieck, "On numerical integration of ordinary differential equations", Math. Comp., 16, 1962, p. 22.

E. J. Nystrom, "Uber die numerische Integration von Differentialgleichungen", Acta Soc. Sci. Fenn., 50, 1925, No. 13.

A. Okubo, Diffusion and Ecological Problems: Mathematical Models, Springer, Berlin, 1980, Chap. 7.

J. M. Ortega and W. C. Rheinboldt, Iterative Solutions of Nonlinear Equations in Several Variables, Academic Press, New York, 1970.

L. B. Rall, Computational Solution of Nonlinear Operator Equations, Wiley, New York, 1969.

E. Rauch, "Discrete, amorphous physical models", Int. J. Theor. Phys., 42, 2003, p. 329.

J. R. Rice, "Split Runge–Kutta methods for simultaneous equations", J. Res. NBS, 64B, 1960, p. 151.

R. D. Richtmyer and K. W. Morton, Difference Methods for Initial Value Problems, 2nd Ed., Wiley, New York, 1967.

S. M. Roberts and J. S. Shapiro, Two-point Boundary Value Problems: Shooting Methods, Elsevier, New York, 1972.

S. L. Ross, Differential Equations, 3rd Ed., Wiley, New York, 1984.

W. Rumelin, "Numerical treatment of stochastic differential equations", SIAM J. Num. Anal., 19, 1982, p. 604.

C. Runge, "Uber die numerische Auflosung von Differentialgleichungen", Math. Ann., 46, 1895, p. 167.

H. E. Salzer, "Oscillatory extrapolation and a new method for numerical integration of differential equations", J. Franklin Inst., 262, 1956, p. 111.

A. M. Samoilenko and N. I. Ronto, Numerical-Analytical Methods of Investigating Periodic Solutions, MIR Pub., Moscow, 1979.

D. Sarafyan, "7th-order 10-stage Runge–Kutta formulas", TR #38, Math. Dept., LSU in New Orleans, 1970.

D. Sarafyan, "Continuous approximate solution of ordinary differential equations and their systems" Comp. Math. Applic., 10, 1984, p. 139.

D. Sarafyan, "New algorithms for the continuous approximate solution of ordinary differential equations and the upgrading of the order of the processes", Comp. Math. Applic., 20, 1990, p. 77.

H. Schlichting, Boundary Layer Theory, McGraw-Hill, New York, 1960.

R. E. Scraton, "Estimation of the truncation error in Runge–Kutta and allied processes", Comp. J., 1964, p. 246.

E. B. Shanks, "Solution of differential equations by evaluations of functions", Math. Comp., 20, 1966, p. 21.

M. Shillor (Ed.), Recent Advances in Contact Mechanics, Special Issue, Math. Comput. Modelling, 28, 1988.

G. S. Simmons, Differential Equations, McGraw-Hill, New York, 1972.

J. Spanier and K. B. Oldham, An Atlas of Functions, Hemisphere, New York, 1987.

M. S. Steinberg, "Reconstruction of tissues by dissociated cells", in Models for Cell Rearrangement, G. D. Mostow, Ed., Yale University Press, New Haven, 1975, p. 82.

G. Strang and G. J. Fix, An Analysis of the Finite Element Method, Prentice-Hall, Englewood Cliffs, 1973.

M. Toda, "Wave propagation in an harmonic lattice", J. Phys. Soc. Japan, 23, 1967, p. 501.

M. Urabe, "On a method to compute periodic solutions of the general autonomous system", J. Sci. Hiroshima University, Ser. A, 24, 1960, p. 189.

M. Urabe, "Theory of errors in numerical integration of ordinary differential equations", J. Sci. Hiroshima University, Ser. A-1, 25, 1961, p. 3.

R. A. Usmani, "A method of high-order accuracy for the numerical integration of boundary value problems", BIT, 13, 1973, p. 458.

B. van der Pol, "The nonlinear theory of electric oscillations", Proc. IRE, 22, 1934, p. 1051.

T. von Karman, Aerodynamics, McGraw-Hill, New York, 1962.

G. Wang and M. Ostoja-Starzewski, "Particle modeling of dynamic fragmentation: I. Theoretical considerations", Comput. Mater. Sci., 33, 2005, p. 429.

J. Weissinger, "Numerische Integration impliziter Differentialgleichungen", ZAMM, 33, 1953, p. 63.

E. T. Whitaker and G. N. Watson, Modern Analysis, Cambridge University Press, Cambridge, 1952.

D. M. Young and R. T. Gregory, A Survey of Numerical Methods, Addison-Wesley, Reading, 1973.

Index

a
air, homogeneous   62
Airy function   155
algebraic system   133, 146
    tridiagonal   133
Algorithm
    Euler   4
    Kutta for second order equations   28
    Kutta systems   25
    leap frog   53
    linear boundary value problem
        solution   146
    Newton–Lieberstein   138
    stable   130
    Taylor   39
    Taylor system   41
    top   103
angstrom   51
argon   50
asymptotically stable   128
autonomous system   122
average velocity   56

b
Boltzmann constant   52
boundary value problem   143, 177
    mildly nonlinear   150
Brownian motion   56

c
Calogero system   103
cavity problem   54, 63
cellular self reorganization   73
chaos   184
classical molecular potential   50
conservation
    angular momentum   81
    energy   79
    linear momentum   80
conservative methodology   77
consistency   181
contact mechanics   184

convergence theory   5
covariance   82, 163, 170
    rotation   82
    translation   82
    uniform motion   82
critical point   122

d
difference approximation   144
difference equation   3, 90
difference quotient   2
differental system
    first order   23, 37
    large second order   49
    second order   26, 41
differential eigenvalue problem   182
differential equation   1
direct method, linear algebraic system   136
discrete function   3

e
eigenvalue problem   182
Einstein   164
energy   79, 118
Euler's Method   1

f
first order system   90
force   29, 31
    impulsive   31

g
gravitation   109
grid point   2, 55
grid size   2
gyroscope   108

h
harmonic oscillator   109
    relativistic   169
homogeneous air   62

i

impulsive force   31
inertial frame   160
initial condition   1, 29
initial data   55, 64
initial value problem   1, 31, 37
instability   111

k

Kutta's formulas   23, 26, 31

l

lab frame   160
leap frog formulas   52
Lennard–Jones potential   50, 51
linear algebraic system   133, 146
linearization   30
Lorentz transformation   161

m

matrix operations   187
microdrop collision   70
mildly nonlinear boundary value problem   150
mildly nonlinear system   127, 138, 150
molecular mechanics   52
molecular potential   50
molecule   50
multistep method   180

n

N-body problem   49
nano system   49
Newton's dynamical equation   29
Newton–Lieberstein method   137
Newtonian iteration method   78
nonlinear pendulum   28
numerical solution   6

p

pendulum   28
periodic solution   43
potential   50
predictor-corrector formula   179
primary vortex   56, 65

r

regular triangular grid   55
relativistic harmonic oscillator   169
relativity   159
rest mass   164

rocket frame   160
rod contraction   161
roundoff error   8
Runge–Kutta formula
    eigth order   18
    fifth order   16
    fourth order   16, 23
    second order   13
    seventh order   17
    sixth order   17
    tenth order   19
    third order   15

s

self reorganization   73
small vortex   59, 60, 68
special relativity   159
spinning top   85
stable   123, 124
stiffness   45
systems
    Calogero   103
    first order   23, 37
    large second order   49
    second order   26, 41
    Toda   103

t

Taylor expansion formula   12, 13, 37
Taylor expansion method   12, 13, 37
Taylor series formula   12, 13, 37
temperature   52
time dilation   161
top, spinning   85
tridiagonal algebraic system   133
turbulent flow   59, 66

u

unstable   111, 124
upwind differencing   148

v

van der Pol oscillator   43
vortex
    primary   56, 65
    small   59, 60, 68

w

wall reflection   56
water vapor   70